SYSTEMS ANALYSIS FOR
PRODUCTION OPERATIONS

STUDIES IN OPERATIONS RESEARCH

Editor: A. Ghosal (Formerly edited by Burton V. Dean)

SYSTEMS ANALYSIS FOR PRODUCTION OPERATIONS

C. Carl Pegels

Associate Professor of Management Science
State University of New York at Buffalo
Buffalo, New York

Gordon and Breach Science Publishers
New York   London   Paris

Copyright © 1976 by

Gordon and Breach Science Publishers Inc.
One Park Avenue
New York, N. Y. 10016

Editorial office for the United Kingdom

Gordon and Breach Science Publishers Ltd.
41/42 William IV Street
London W.C.2

Editorial office for France

Gordon & Breach
7-9 rue Emile Dubois
Paris 75014

Library of Congress catalog card number 75-29596
ISBN 0 677 04710 X

PREFACE

This book is intended to provide the reader with an appreciation and understanding of the process and quantitative tools of systems analysis for production operations. It is intended as a reference book for practitioners and as a text book for undergraduate and graduate courses in business administration, management science, operations research, management and industrial engineering which deal with systems analysis, analytical planning, and decision making.

Although the material presented in this book has been around for some time it is generally recognized that the application of quantitative techniques such as those presented here will become more common in the future. Robert F. Vandell (Harvard Business Review, January-February 1970, p. 83) in his article "Management Evolution in the Quantitative World" predicts that the combination of quantitative analysis and computer availability constitute a powerful force which will force management to undergo some radical changes. He points out that most of the techniques that will be needed have already been developed. The developers of these techniques have usually been from the academic world with little interest in testing out these techniques in real applications. Management, on the other hand, has been slow in taking advantage of these potential management tools mainly because of lack of adequately trained manpower to use the developed tools on practical problems or for decision making. What becomes evident is that there is a tremendous gap between theory and application. This book aims to narrow the gap between theory and application. Traditional operations researchers have been trained to work on theoretical projects and hence have had difficulty to adapt their knowledge and expertise to real problems. Increasingly, management is channeling the operations research efforts towards real, smaller and more manageable projects and the results are becoming evident in many firms.

Why has there not been more progress in the adoption of quantitative techniques to help firms solve problems and make decisions? As Vandell points out manpower has been short. At present there are only about 20,000 qualified people with a deep interest in management science who also have the professional competence to apply sophisticated operations research tools to complex operations decisions. Of the 20,000 only about 4000 are actually working in business and industry; the remainder are academically oriented or work with computer firms. However, the large number of management scientists in academia has created a large training resource for operations researchers and as many as 10,000 a year will be entering business and industry during the mid-seventies. This book will contribute to close the gap between theory and application for these trainees.

The material covered in this book is a compilation of techniques and applications of techniques developed during recent years. Numerous books have been written on production management, operations management, operations analysis, management science or applied operations research. These books have frequently attempted to cover one or another aspect of systems analysis for production operations but because of their breadth of coverage have never been able to cover it thoroughly.

An extensive amount of material is included on production facilities planning. Facilities planning is an activity that occurs only once in the lifetime of a production, distribution or service facility. Operations planning on the other hand takes place continually during the lifetime of a facility. As a result it is understandable why text and reference books on operations planning are found in abundance. However, although facilities planning only takes place once in the lifetime of a facility, it is extremely important because it can significantly affect the cost of operating the facility during its lifetime. Therefore, anyone participating in facilities planning, design, evaluation, analysis and decision making should be aware of the essential models described in this book. Most facilities planning decisions are irreversible. Utmost care should therefore be taken that only the best decisions are made. It is the aim of this book therefore to assist in making only the best facilities planning decisions.

The first part deals with systems analysis related to facilities planning and the tools of systems analysis. Simulation, one of the main tools, is discussed in considerable detail. Systems analysis applications in divergent cases are presented. Part two discusses capacity planning models. A more theoretical approach is taken here illustrated with several real life applications. Part three deals with layout planning models, the physical part of facilities planning. The reader will find that in this area significant payoff can be realized by detailed analysis using the available tools. The final part discusses several operations planning models as they relate to facilities planning. The reader who wants to expand his knowledge on any of the models will find suggestions for further study in the references which are all listed in the bibliography.

Acknowledgements are due to the secretarial staff of the School of Management at the State University of New York at Buffalo and to students who used mimeographed versions of much of the material in this book. They were excellent critics. The various publications that allowed me to use previously-published material are also due a special note of thanks. Thanks are also due those authors from whose work I have learned and whose ideas will have penetrated the material herein presented. I have attempted to always give credit where credit was due. If I have failed anywhere please accept my apologies.

C. Carl Pegels

LIST OF ILLUSTRATIONS

x

TABLE OF CONTENTS

PART I.--SYSTEMS ANALYSIS AND FACILITIES PLANNING MODELS

PART IV.--OPERATIONS PLANNING MODELS

Hiring and Layoff Costs. Overtime and Idle Time Costs.
Inventory and Shortage Costs. GENERAL MODEL. Derivation of Decision Rules. Decision Rules for Production
and Work Force. Analysis of Decision Rules. THE JOB
SHOP OPERATION. The Cost Model for the Job Shop Operation. APPLICATION OF MODEL. Effectiveness of Model
Application. EXERCISES.

PART I. SYSTEMS ANALYSIS AND
FACILITIES PLANNING MODELS

In the first chapter systems, systems analysis, simulation, and modeling will be discussed as an introduction to several systems studies. Two short and rather simple simulations of decision problems will be presented to illustrate how simulation can aid the decision process.

The simulation examples at the end of the first chapter also serve as an introduction to a considerably more complex simulation study in the second chapter. The more complex simulation study illustrates how a complex industrial system can be sufficiently simplified so that it can be simulated, and still satisfy a rigorous validation test.

In the third chapter the validated simulation model will be used to test out numerous alterations in the operation and physical dimensions of the industrial system. In other words the simulation model will be used to experiment with the industrial system without disturbing the system. The only cost of this experimentation, once the simulation model is built, is the computer time cost.

Simulation is used extensively in industry, business and non-profit institutions to model the entity studied and to determine the effects of "what if" policies. During the past several years simulation has become an important managerial tool for planning and forecasting in organizations and for evaluating the operations in these organizations. In a book edited by Schrieber[1] several applications of simulation to organizations are described.

---

[1]Schrieber, A.N., _Corporate Simulation Models_, Graduate School of Business Administration, University of Washington, 1970.

The volume consists of papers presented at a symposium on simulation models and examples of simulation applications are presented in firms engaged in the manufacture of airplanes, computers and glass. Other firms whose operations have been simulated are in banking, chemicals, farm equipment, oil and forest products to name just a few.

Application of the simulation approach to a machine tool manufacturing firm is described by Desjardins and Lee in the previously referenced book by Schrieber. The simulation model developed for this firm was designed to simulate the physical flow of materials, cash, and information within the firm. A modular approach was used to develop the model. The purpose of the model is to have a tool for investigating alternative policies the firm may consider.

In addition to the applications presented in Chapters 1, 2, and 3 the author has been engaged in simulating an aerospace job shop, blood banks and blood inventory behavior in large geographic areas. Simulation is currently also applied to study the effects of pollution and pollution abatement policies on waterways and other bodies of water.

The fourth chapter presents a study of an industrial plant's power system. Because this problem is simpler than the problems discussed in the previous chapter, analytic techniques instead of simulation have been used to solve and analyze it.

The techniques used to study the power plant problem, such as discounted cash flow, cost benefit analysis and estimation of distribution functions, are quite common and extensively used by analysts trained in these techniques. Because they are quite common, little is published to illustrate how firms or organizations have used these techniques.

The use of decision trees and Bayesian analysis form the fifth chapter. Three capital investment decision analysis problems are presented and analyzed using decision tree analysis. Decision tree analysis is extensively used in industry as reported by Brown[2]. He reports that Dupont has used this technique since the early 1950's and Pillsbury has used it since the early 1960's. Since 1964 interest in the technique has been stimulated and numerous other well-known companies have begun using decision tree analysis. For instance, General Electric made a study of the technique that led to major changes in plant appropriation methods; Ford Motor trained hundreds of their middle and senior managers in applying the technique; and General Mills began to introduce the technique on a project-by-project basis.

The sixth and last chapter of the first part is also on the subject of capital investment decisions. However, simulation is the vehicle for analysis and several capital investment decision criteria are evaluated. The first successful application of capital investment simulation was reported by Hess and Quigley.[3] They analyzed rate of return and cash flow resulting from

---

[2] Brown, R. V., "Do Managers Find Decision Theory Useful?" Harvard Business Review, May-June 1970, p. 78.

[3] Hess, S. W. and H. A. Quigley, "Analysis of Risk in Investments Using Monte Carlo Techniques," Chemical Engineering Progress Symposium Series 42: Statistics and Numerical Methods in Chemical Engineering, New York: American Institute of Chemical Engineering, 1963, p. 55.

an investment in a chemical process. Hertz[4] subsequently refined the technique and applied it to a $10 million extension of the processing plant of a medium size industrial chemical producer. The application revealed that the technique provides management with far more information on which to base a decision than on an expected return type of decision which they previously had used.

Carter[5] reports that firms in the oil industry, which have management people with adequate training in management science, have little difficulty in adopting the techniques described in Chapters 5 and 6. However, he also points out that failures at using management science techniques may be expected if management does not have adequate background in using these techniques.

---

[4]Hertz, D. B., "Risk Analysis in Capital Investment," Harvard Business Review, January-February 1964, p. 95.

[5]Carter, E. E., "What are the Risks in Risk Analysis," Harvard Business Review, July-August 1972, p. 72.

CHAPTER 1.--INTRODUCTION TO SYSTEMS ANALYSIS

This chapter is intended to provide the reader with
some background in systems, systems analysis, simulation and
modeling in order that he may appreciate the subsequent
chapters where a certain amount of proficiency is assumed.
Also at the end of this chapter two rather straightforward
decision problems are presented to illustrate systems analysis and simulation. Simulation is used as a technique to
shed more light on the dynamics of these two problems and
the decision maker thus has considerably more information
on which to base his decisions. What the simulation examples make perfectly clear is that all activities that are
known by such names as systems analysis, simulation, etc.
really only are a means for providing additional information
to the decision maker but do not make the decisions. It is
the decision maker who must decide which alternative or alternatives to select.

SOME DEFINITIONS

Numerous terms are used in the study of systems and
simulation which may have more than one meaning, or may mean
one thing to one reader and something else to another reader. An attempt will therefore be made to define as many
terms as possible at the outset.

Simulation, Experiment and Model

In order to have a starting point from which other
terms may also be introduced we shall define the word simulation first. Simulation is the building of an abstract
model of a system and performing experiments on the model.
This definition requires a defining of the terms experiment,
model, and system.[1]

An experiment will be defined as operating the model
under specified input conditions and observing the performance and output of the model. This term is general enough
so that it may apply to any collection of data. For instance the taking of a roll call is in essence an experiment, and so is weighing yourself on a bathroom scale.
However, for simulation purposes the more restrictive form
defined above will be used.

The term model or at least the concept of a model
has been around for centuries. Automobile and aircraft
firms have been making "mock-ups" or models of proposed
future products for a long time. Similarly, engineers engaged in new plant design have been making models or layouts
of proposed locations of the various pieces of equipment.

---

[1]See also J.H. Mize and J.G. Cox: Essentials of Simulation. Englewood Cliffs, N.J.: Prentice-Hall, Inc., 1968,
p. 1.; and R.C. Meier, W.T. Newell, and H.L. Pazer: Simulation in Business and Economics, Englewood Cliffs, N.J.:
Prentice-Hall Inc., 1969, p. 1.

The type of models to be discussed here are analogous to the classical models discussed above. However, the distinct difference is that the classical models are physically concrete, whereas the models discussed here are abstract. An abstract model is a representation of components of the real system which will affect the behavior of the system. A simple example of an abstract model is a firm's income statement. The income statement represents certain quantitative attributes of the firm's operating system.

As we stated before the hazard of defining terms results in the need to define more terms. Several new terms have been injected into the above discussion without proper definition. This will be corrected with a detailed definition of a system, its components and attributes.

Systems

A system will be defined as a group of components which do or could interact with each other. Because of the possibility of interaction some of the components may be dependent on other components, or interdependencies may exist among components.

Each component of a system may have certain properties called attributes. The number of attributes of a component of a system is usually very large. However, only those attributes relevant to the study of the system are considered in a systems analysis.[2]

Some of the attributes of system components are qualitative and others are quantitative. The quantitative attributes are especially important because they are so much easier to manipulate. In a complex system a group of observed or dependent variables y_i is related or assumed to be related to another group of known or unknown independent variables x_i. If we let y be the vector of dependent variables y_i, and x be the vector of independent variables x_i, then we can express the relationship between the two vectors as

$$y = g(x)$$

where g is generally a non-linear, complex and unknown function.

SYSTEMS ANALYSIS

Let us now take a look at systems analysis and simulation. Systems may be analyzed by means of a simulation model. However, simulation is not the only means of analyzing a system. As a matter of fact, it is frequently argued that simulation should always be the last resort and the alternatives of analytic methods or direct experimentation

---

[2]See also C. McMillan and R.F. Gonzalez: Systems Analysis, Homewood, Illinois: Richard D. Irwin, Inc. 1968, pp. 1-12; and K.D. Tocher: The Art of Simulation, Princeton, N.J.: D. Van Nostrand Co., Inc., 1963.

should always be considered first. The reason simulation is so extensively used is because systems are often too complex to allow the use of analytic techniques and direct experimentation is seldom feasible.

There are several phases to a systems analysis. These are generally considered to consist of:

1. formulation of the problem

2. design of the model

3. derivation of a solution from the model by simulation or analytic techniques

4. validating the model and making adjustments if required

5. experimentation and/or derivation of a solution

The first item is important to ensure that the correct model is designed. As was pointed out above only certain attributes of system components are involved in any systems analysis. Therefore, the model designer must include all the pertinent ones and leave out those attributes which do not affect the solution. This process is not as simple as it sounds, and frequently models require redesign if it is found impossible to obtain a good validation of the model. The last item indicates that most systems analyses provide more than just one solution. That is, the original objective of most systems analyses is the acquisition of data on the system's behavior under various **external** or internal conditions.

Although, simulation is the most common method for analyzing systems, especially complex systems, <u>analytic techniques</u> are frequently used provided the necessary simplifying assumptions do not produce a model which deviates too far from reality. For example, in the third chapter simulation results together with analytic methods are used to find answers to a capacity problem. In chapter four analytic techniques are used to find the optimum level of the power factor in a plant power system. As a matter of fact, in most of the remaining chapters of this book analytic techniques are used to derive solutions or if you like to analyze the system after the problem has been formulated. From the description of the system analysis techniques stated above you may thus note that this book is a book on systems analysis applications.

Since simulation will be the vehicle used for a complex systems analysis study in the next two chapters a little more space will be devoted to the subject of simulation in the remainder of this chapter. Two simulation examples will also be presented to illustrate applications of the simulation approach.

Simulation

Simulation is like a model or "mock-up". Automobile and aircraft firms make "mock-ups" of proposed future products and study how these "mock-ups" behave under various

external and internal conditions. Of course, a physical
model of a plant can be made, but it would not be very useful except for decorative and plant layout purposes. What
is useful, however, is a model of a process, a firm or plant
that incorporates relations which affect output (profit, revenue, time taken, etc.) like production functions, cost
functions, machine breakdown functions, labor contract parameters, and so forth.

Shubik defined simulation as follows. "A simulation
of a system or organism is the operation of a model or sim-
ulator which is a representation of the system or organism.
The model is amenable to manipulations which would be impossible, too expensive or impractical to perform on the
entity it portrays."[3]

Simulation is often criticized and called an inferior method to find solutions to complex problems. It is,
however, an engineering idea, analogous to the pilot plant,
towing tank and wind tunnel. As Bellman has stated, "The
use of simulation techniques represents the application of
standard scientific methods to areas which have not previously been treated to any extent by scientific techniques."[4]

---

[3]M. Shubik, "Simulation of the Industry and the Firm,"
American Economic Review, Vol. 50 (December 1960), p. 909.

[4]R. Bellman, "Top Management Decision and Simulation
Processes," Journal of Industrial Engineering, Vol. 9
(September-October 1958), p. 461.

Before a number of advantages of simulation are listed, it will again be stressed that simulation methods should only be used if analytic techniques are infeasible or impossible, or it may be used as an additional way to evaluate a problem situation. The many examples of applications of analytic techniques presented in this book point out that simulation is not always the only alternative. However, simulation is often viewed as the only alternative to study a system. The explanation of this viewpoint may be contained in the list of simulation advantages cited by Spencer.

1. "A computer model or simulation can be much more realistic than a traditional mathematical type of model. This is because the latter, requiring an analytic solution, must of necessity be kept simple, if not overly simple. Computer (simulation) models on the other hand, are solved by tracing the time paths of the endogenous variables, i.e. by simulating the process contained in the assumptions; hence, they may be as complex and realistic as the underlying theories require. A chief advantage of computer simulations, therefore, is that they provide a concrete procedure for formulating and testing hypotheses."

2. "The simulation process makes it possible to conduct analyses of a model by running the model numerous times. In this manner, one or more assumptions, relations, or variables can be changed or modified as necessary, with

much greater ease than would be possible if analytic solutions were sought."

3. "Simulation permits the varying of "time" in
order to handle objects which move at different rates of
speed, and it observes long periods of operations in mat-
ters of minutes, thus permitting the foreseeing of problems before they develop. The dynamics of complex systems
are thereby revealed in a manner not hitherto possible."

4. "Computer models can be employed by nonmathe-
matical analysts, thereby widening their effectiveness in
the study of complex social systems."

5. "The simulation process permits the large-scale
testing of nonlinear problems. In addition, it often aids
the analyst in gathering valuable information about the system under investigation, and in many instances is more
economical to use than other experimental processes."[5]

The disadvantages of computer simulation are many,
although not as obvious as the advantages. Because a computer simulation can deal with rather complex relationships,
it is frequently difficult to determine the number of simplifying assumptions to use. In an analytical model, the
simplifying assumptions are mandatory and a function of
the analytic model being used. In simulation the degree of

---

[5] M.H. Spencer, Managerial Economics. Homewood, Illinois:
Richard D. Irwin, Inc. 1968, p. 454.

simplification is determined by the model builder. The result is that simulation models are frequently made too complex which causes problems with the validation of the model.

Validation

A computer simulation model of a system should be validated. A validation consists of comparing the model's performance with the performance of the actual system being simulated. Both input and output variables of the simulation model should reasonably replicate the actual system. What is reasonable is a difficult question by itself. In chapter two the difficult validation problem will be reviewed in more detail.

Finally, there is the very difficult problem of determining which variables to include in the simulation, or rather which variables not to include. Excessive complexity is not desired, but sufficient detail should be included to ensure a valid model can be obtained. Related to the problem of which variables to include is the added problem of obtaining parameter estimates to be used as inputs to the system. This problem is especially difficult because the observations on data used for parameter estimates must frequently be modified to take into account the necessary simplifying assumptions introduced at other points of the model design. Hence, counterbalanced against Spencer's simulation model advantages are just some of the many disadvantages discussed above.

SIMULATION EXAMPLES

Two examples of the use of simulation will be presented to provide the reader with some appreciation of the way a simulation of a process illustrates the process behavior over time. Most analytic solutions only provide a snapshot kind of solution. Simulations on the other hand present a picture of the behavior of the process over time analogous to a movie.

The first example is a periodic investment problem under risk. The probability of a large payoff is small and the probability of no payoff is considerable. The expected payoff, however, is substantial in relation to the cost of participating. The second example analyzes the classical paper boy problem. How many papers should the paper boy buy in order to maximize his profit over time?

Periodic Investment Problem

This problem can also be viewed as the risk problem of having to invest a large sum of money to obtain a still larger payoff with probability p and no payoff with probability 1-p. This problem applies to the speculator, the oil well driller and to anyone who has frequent investment opportunities to consider.

The problem is essentially a gambler's problem with the exception that the gambler's net expected payoff is usually less than his investment. The oil well driller, on the other hand, will never consider drilling an oil well with a cost greater than the expected payoff. Similarly, a professional speculator will not speculate unless the expected revenue exceeds the cost of the speculation.

Suppose it costs $100 to get a chance at receiving a return of $1500 with a probability of $\frac{1}{6}$, and receiving nothing with a probability of $\frac{5}{6}$. The expected value of the speculation, E, is $1500 $(\frac{1}{6})$ + $0 $(\frac{5}{6})$ = $250. Hence, the payoff is positive. There are, however, some constraints on the speculator, or rather the entrepreneur. He can only make one bet during each period, and he must borrow his money at a capital cost of 10% per period. The question that the entrepreneur would like an answer to is if the investment or gamble each period is still a worthwhile endeavor. Since he starts out with borrowed money he must continue to borrow until he wins a sufficient number of times, so that he can repay his debt plus accumulated interest charges.

A simple simulation will be run for the entrepreneur which will show him what he can expect to see over time if he decides to accept the challenge. To generate probabilities of $\frac{1}{6}$ a die can be used and if the total

eyes on the die total five the entrepreneur receives a payoff of $1500. If any other number comes up, he will lose
his investment. In Table 1-1 the reader can follow the
entrepreneur's simulated experience.

Note that the entrepreneur's net gain after twenty-
four periods is $1746.40. Had he been successful earlier
his total payoff could have amounted to well over $3500.
The simulated problem is a rather simple one yet is complicated enough to benefit from a simulation model approach to
illustrate the behavior of the stochastic cash flow over time.

This problem could have been solved by direct analytic methods. However, only the expected payoffs would have
been obtained and a picture of the possible dynamic behavior
of the cash position of the entrepreneur over time would not
have been available. What this suggests, however, is that
a simulation provides a desirable additional means of evaluating a decision problem even though it has already been
analyzed by analytic methods.

Paper Boy Problem

A paper boy is faced with the problem of having to
decide how many papers he should order daily in order to
maximize his average daily profit over time. He purchases
papers for $0.05 per copy and sells them for $0.10. Any
unsold papers are worthless. He knows that the demand is

Table 1-1

Simulation Results of Entrepreneur's Speculation

Period	Investment	Cumulative Investment	Periodic Interest Charge	Payoff	Cumulative Assets	Liabilities
1	$100	$ 100.00	$ 10.00	$ 0		$110.00
2	100	210.00	21.00	0		231.00
3	100	331.00	33.10	0		364.10
4	100	464.10	46.41	0		510.51
5	100	610.51	61.05	0		671.56
6	100	771.56	77.16	0		848.72
7	100	948.72	94.87	0		1043.59
8	100	1143.59	114.36	0		1257.95
9	100	1357.95	135.80	0		1493.75
10	100	1593.75	159.38	0		1753.13
11	100	1853.13	185.31	1500		538.44
12	100	638.44	63.84	0		702.28
13	100	802.28	80.23	0		882.51
14	100	982.51	98.25	0		1080.76
15	100	1180.76	118.08	0		1298.84
16	100	1398.84	139.88	0		1538.72
17	100	1638.72	163.87	1500		302.59
18	100	402.59	40.26	0		442.85
19	100	542.85	54.29	0		597.14
20	100	697.14	69.71	0		766.91
21	100	866.91	86.69	1500	$ 546.40	
22	100			0	446.40	
23	100			1500	1846.40	
24	100			0	1746.40	

distributed according to the uniform distribution. However, the number of papers sold daily depends on whether it is dry or wet. When it rains he sells only between 31 and 40 papers and when it is dry he sells between 41 and 50 papers. The probability of a wet day during the Spring season for which he is trying to maximize his profit is 0.40. How many papers should he order daily?

Although the above problem is rather simple it is an interesting problem to solve by simulation. The paper boy's profit function can be stated as follows:

$$\Pi = 0.10s - 0.05p \qquad s \leq p$$

where Π is profit in dollars, s is number of papers sold and p is number of papers purchased.

The simulation involves only a few steps to simulate a day. First it is determined by a random draw from a table of random numbers[6] whether it is wet or dry. The numbers between 0 and 39 indicate rain and between 40 and 99 indicate dry weather. If it rains a random number between 31 and 40 is drawn from a table of random numbers to determine the number of papers sold that day. Similarly if it is dry a random number between 41 and 50 is drawn. After the sales for the day have been determined the profit

---

[6]Tables of random numbers may be found in many text-books on simulation and in many handbooks.

is calculated with the profit function for a given number of papers purchased. Suppose the calculation is made for 36 to 45 papers purchased in increments of one. The flow chart in Figure 1-1 illustrates the steps involved in a one-day simulation.

Table 1-2 presents the results of the simulation for a twenty-five day period. For illustrative purposes only twenty-five days were simulated. However, if one programs it for a computer it is desirable to increase the number of simulation runs to ensure that a representative sample has been drawn from the respective distributions. For the twenty-five day period the optimum number of papers the paper boy should purchase is 42. This will give him an average daily profit of $1.896. Keep in mind, however, that the simulation was only for the Spring with its own peculiar demand which is dependent on the weather. Another simulation with different parameters would have to be executed for other seasons.

The above problem, as well as the previous example, can be evaluated analytically. However, the simulation presents a typical pattern of how daily sales and profit behave over time. This pattern, together with the analytic solution will aid him in the decision process of selecting the best alternative.

In the next chapter a simulation model will be presented of a considerably more complex operation where simple

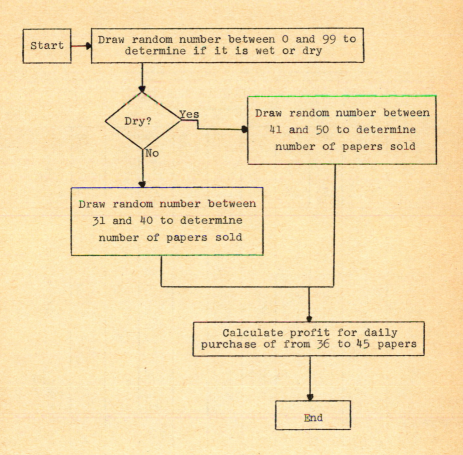

Figure 1-1

Flow Chart of Paper Boy Problem

Table 1-2

Paper Boy Simulation Results

Papers Sold	Profit as a function of number of papers purchased in cents									
	36	37	38	39	40	41	42	43	44	45
41	180	185	190	195	200	205	200	195	190	185
40	180	185	190	195	200	195	190	185	180	175
42	180	185	190	195	200	205	210	205	200	195
48	180	185	190	195	200	205	210	215	220	225
45	180	185	190	195	200	205	210	215	220	225
37	180	185	180	175	170	165	160	155	150	145
49	180	185	190	195	200	205	210	215	220	225
50	180	185	190	195	200	205	210	215	220	225
41	180	185	190	195	200	205	200	195	190	185
48	180	185	190	195	200	205	210	215	220	225
38	180	185	190	185	180	175	170	165	160	155
50	180	185	190	195	200	205	210	215	220	225
47	180	185	190	195	200	205	210	215	220	225
42	180	185	190	195	200	205	210	205	200	195
42	180	185	190	195	200	205	210	205	200	195
47	180	185	190	195	200	205	210	215	220	225
49	180	185	190	195	200	205	210	215	220	225
44	180	185	190	195	200	205	210	215	220	225
36	180	175	170	165	160	155	150	145	140	135
32	140	135	130	125	120	115	110	105	100	95
39	180	185	190	195	190	185	180	175	170	165
36	180	175	170	165	160	155	150	145	140	135
34	180	155	150	145	140	135	130	125	120	115
46	180	185	190	195	200	205	210	215	220	225
37	180	185	180	175	170	165	160	155	150	145
Total Profit	4440	4525	4590	4645	4690	4725	4740	4725	4710	4685
Average Daily Profit	177.6	181.0	183.6	185.8	187.6	189.0	189.6	189.0	188.4	187.4

approaches such as described for the above illustrative examples are not sufficient to resolve the problem.

EXERCISES

1. Solve the entrepreneur's problem (first simulation
    example) by analytic (probabilistic and mathematical)
    methods.

2. Solve the paper boy problem by analytic methods. How
    does it compare with the simulation solution?

3. Simulate a three seat barber shop. Assume customers
    arrive at the rate of 0, 1 or 2 every five minutes,
    i.e. 0, 1 or 2 customers with probability of $\frac{1}{3}$ each.
    Each chair services a customer in 10 to 20 minutes.
    Assume that the time a customer spends in the chair
    is uniformly distributed. Simulate the barbershop in
    five minute increments and determine the longest wait-
    ing time, average waiting time, largest number of cus-
    tomers waiting, etc. Run the simulation for a four or
    five hour period.

4. Simulate the above barbershop with four seats and eval-
    uate the new arrangement. How much idle time is in-
    curred by the barbers?

5. How much should the average charge per customer be in
    order to give each barber an approximate salary of
    $200 per week with a) three seat shop, and b) a four

seat shop? Assume that the rental and maintenance of the shop amounted to $150 per week. The barbershop is open 54 hours per week.

6. The management of a distributor of cranes used in construction is concerned with adjusting its inventory to yield maximum profit. The current mode of operation is as follows:

(a) A unit is ordered from the factory as soon as one is sold, so that the sum of units on hand plus the unfilled replacement orders is a constant K.

(b) Sales have a Poisson distribution with rate λ/week.

(c) Prospective customers are lost whenever inventory is zero.

(d) Lead time for receiving an item, which has been ordered is exponentially distributed with mean replenishment time T.

For,

C_1 = sales price

G = Gross profit/unit sold = sales price - cost price

C_I = inventory cost/unit/unit time

C_p = cost per replenishment order,

solve for K using a simulation approach and reasonable parameter values. Is there and analytic solution to the problem?

7. Run a simulation for the following toll barrier problem. Automobiles must stop to pay a 15¢ toll at the toll barrier. There are three gates open; lane one is an attended booth for customers who do not have the correct change; and lanes two and three are automatic booths for customers who have the correct change.

   The time between arrivals of customers is uniformly distributed on the interval two to six seconds. Gates two and three have a constant service time of six seconds. Eighty five percent of the customers have the correct change.

   A queue for service forms at each of the lanes. A customer arriving with the correct change will select and enter the line which is shortest at the time he arrives. In the event when both lines are equal, he will select either lane with probability of 0.5. Find what is:

   (a) percentage utilization of each lane.

   (b) average time a customer waits in each of the queues.

   (c) Maximum and average queue length of each of the queues.

8. A supermarket is concerned with the possibility of adding an express check-out service (for shoppers with five or fewer items). An analysis of store traffic reveals that the arrival and service times for this

type of shoppers are as follows:

Number of Arrivals per minute	Probability	Service time Minutes	Probability
0	0.80	1	.70
1	.10	2	.25
2	.05	3	.05
3	.03		1.00
4	.01		
5	.01		
	1.00		

(a) Using a simulation approach, calculate the average length of the waiting line, average waiting time of customers, average time per customer and percent idle time of the checker.

(b) Should the facility be opened, if the average idle time of the checker is greater than fifty percent?

9. To be operational, a system requires that a particular component be on line and in working condition. Once the system is operational, the time to failure is uniformly distributed between 80 and 120 hours. The time to repair the component, off line, is uniformly distributed 70 to 90 **hours. Only** one component can be in the repair facility at a time. To install a component requires two hours. No time is required to remove a defective component.

How many of these components should the system have available to be operational at least 90% of the time?

10. Consider the following inventory system for maintaining spare parts for a certain component in a system. When the quantity of parts falls to or below the re-order point R, a replenishment order for Q units is placed. These units are received after a lead time which is uniformly distributed 20-40 days. Failures to this component in the system which require the replacement of the failed part by one from inventory occur in a random fashion. Examination of the historical data has shown that the time between failure is exponentially distributed with a mean of 10 days. A failed part is discarded rather than repaired.

For Q = 6 and R = 4, what percentage of the time is the system inoperative for lack of spare parts?

11. You have been asked to analyze the operation of a hospital pharmacy in which two pharmacists fill prescription orders received from various patient units in the hospital. It is the policy that each order is filled by either one of the pharmacists and then is checked by the other. Orders arrive with a uniformly distributed inter-arrival time between three to six minutes and are placed in a basket at the pharmacy. A pharmacist will pick it up, check it and place it on a shelf where it is picked up by the nurse. A pharmacist will fill a

new order only when he has completed checking all orders in the to-be-checked area by his partner. The time to fill an order is exponentially distributed with a mean of three minutes while the time to check an order is exponentially distributed with a mean of one minute. Determine the system utilization of each pharmacist and the average time for each prescription to be processed.

12. Transamerican Airfreight delivers and picks up airfreight shipments from customers in most of the major cities. Upon pickup of the airfreight, Transamerican pools the airfreight at major airports and dispatches it to the various air carriers depending on destination. The rate of pickup requests and freight arrivals varies considerably from day to day. However, it has been found that pickup requests and arrivals are independent of each other and both are normally distributed with means of 760 and 685 and standard deviations of 75 and 85 respectively.

The number of stops for pickup or delivery a truck can make per eight hour day averages 20 with a standard deviation of 6 and is also normally distributed. Daily operating cost for a truck amounts to $7 per hour with overtime costing 40% more. Delays in pickups are not allowed, that is, a pickup schedule for a given day must be picked up that day. However, a delivery may on occasion be delayed for one day and sometimes for two days. Management does not know, however, how costly delayed shipments are. Transamerican has two competitors in most locations and regular customers could be lost if poor delivery service were provided.

How many trucks should Transamerican have in operation to provide on time delivery of 99% of all shipments?

13. What is the appropriate marginal cost per shipment to Transamerican (see exercise 12) to provide on-time deliveries over and above the average daily cost of $2.80 per shipment? What percentage of shipments would you specify to be delivered on time if you were the president of Transamerican?

14. Appollo Products is a producer of subassemblies for computer peripheral equipment. It also machines nearly all of the component parts that go into these subassemblies. Appollo's machine shop policy on overtime is as follows: no overtime is worked until the backlog workload for any work center exceeds the equivalent of sixteen hours of work. The shop only works one eight hour shift thus a sixteen hour backlog equals two days' work. However, a one day notice is required to workers informing them of the need for overtime. Hence, overtime is always worked a day later than needed. Order arrivals on any day for a workcenter are normally distributed and are measured in hours of work. The mean number of orders arriving daily equals eight with a standard deviation of two orders. Run a simulation for 100 working days and determine how many hours of overtime will be scheduled. If there is no work to be done the work center operator remains on the payroll.

15. In reference to exercise 14, work center operators have indicated that they are willing to go on a nine hour shift with the ninth hour paid at 125% of regular hourly pay. Overtime is currently paid at 150% of regular pay. Run a simulation to determine if it would be advantageous for Appollo to switch to a nine hour shift.

16. Define simulation, experiment and model and give an example which will illustrate each of the three concepts.

17. Simulation as defined in this chapter consists of two distinct stages. Discuss these two stages and draw parallels with a mathematical model.

18. What is meant by systems analysis. Give a brief example of a system and how you would go about analyzing it.

19. Discuss some advantages of simulation over strict mathematical models and over real-life processes.

20. Why is it important to validate a simulation model? Are all models equally easy to validate?

CHAPTER 2.--SIMULATION OF AN INDUSTRIAL SYSTEM FOR
PLANNING PURPOSES[1]

Numerous papers have been written on simulation and
especially on tactical and Monte Carlo simulation. However,
most of these articles have been expository and general.
The few detailed examples which appear in textbooks are
usually simulations of rather simple systems. The model
discussed in this chapter simulates an industrial system
consisting of a production process, the production scheduling process, and the market process.

OVERVIEW OF THE SIMULATION

The production process being simulated is best
adapted to continuous operation, but the market demand calls
for job-shop operation - meaning that there are many small
orders, each of which has unique specifications. These conditions make the scheduling operation and the production
operation quite complex. Due to this complexity, it is vir-
tually impossible to develop useful analytic techniques to
study and analyze the system with the intention of searching
for improved methods of operation. Simulation, on the other
hand, is an ideal tool to analyze operations like the

---

[1]This chapter is based on C. Carl Pegels: "Simulation
and the Optimal Design of a Production Process," The International Journal of Production Research, Vol. 7, No. 3
(1969), pp. 219-231.

complex system presented in this chapter.

Product Description

The product of the industrial system on which this
simulation is based is corrugated paperboard used in the
production of shipping boxes. The performance characteristics of the product, including its thickness, are determined
by the type of fabrication and the raw material composition.
Orders for the product request varying quantities. Hence,
the specifications of an order consist of five parameters:
type of fabrication, raw material, number of items per order, width of the item and length of the item.

The product is available in four types, in ten raw
material compositions, in any number of items and in widths
and lengths that satisfy the width and length constraints
built into the production facility. It is therefore understandable that each order is essentially unique.

The Three Operations

As was mentioned in the introduction, three operations are simulated. These are the market operation which
generates orders with their respective unique specification
parameters, the scheduling operation which organizes the orders into efficient schedule patterns, and the production
operation.

The production operation converts raw material into a continuous strip of finished product. This strip of finished product is then cut into product items of which the dimensions are specified by each individual order. Frequent change-overs are normal, but much of the change-over work is performed during the running of the previous order or set of orders.

The scheduling operation is a rough analogy to the human scheduler it simulates. It develops schedule patterns consisting of one, two or three orders. A maximum of two orders can be scheduled side by side, as can be seen in Figure 2-1. To obtain schedule patterns consisting of two or three orders, limited over-runs and under-runs are allowed. The reason for running two orders side by side is to obtain better width utilization of the production facility.

Market Operation Definitions

Before discussing the market operation, some terms used in that discussion will be defined. A product line is created by combining a type of fabrication with a raw material composition. Since there are four types of fabrication and ten raw material compositions there could be as many as forty product lines. Only twenty-five product lines have been incorporated in the model, and the empirical system has only a few additional ones which are low-volume lines and are scheduled very infrequently. Hence, we have made a

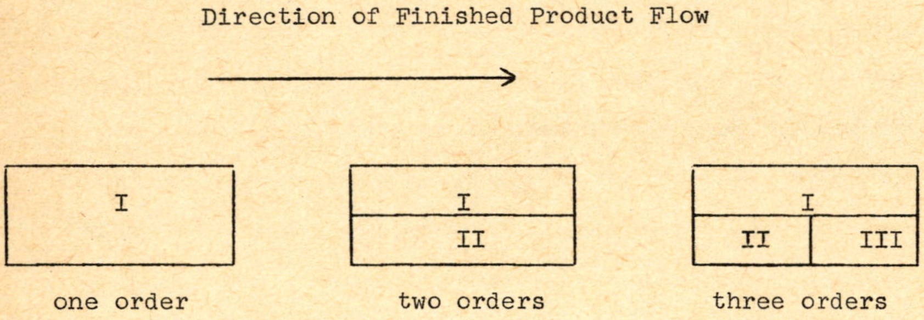
Direction of Finished Product Flow
I
one order
I
II
two orders
I
II
III
three orders

Figure 2-1

Possible Schedule Patterns of Simulation Model

simplifying assumption for which adjustments will be made as will become clear later in the chapter.

A schedule period is determined by the cycle,[2] and therefore these two terms will be discussed together. A cycle is defined as the amount of finished product produced during a time period long enough for the sixteen low-volume product lines to be scheduled once, and the nine, high volume product lines to be scheduled three times. Hence a low-volume product line has only one schedule period during a cycle while the high-volume product lines have three schedule periods. Since the number of orders for each product line, during each schedule period, is randomly determined it is conceivable that a product line may be skipped during a cycle because of lack of orders.

A cycle is a useful, although not necessary construct for a simulation model. Random sampling from an **empirically-** based distribution could also be used to determine the sequence in which product lines are run. However, random sampling would deny the existence of a regularity with which

---

[2]The management of the actual plant knew of no cycle in their operations until the term was introduced by this study. They realized then there was and always had been an approximate cycle in their operations. However, it was not at all as defined as the one finally adopted for the simulation model. The cycle for the simulation model is thus an approximation of the nearly undefinable cycle of the actual operation. In a sense the adoption of a cycle is a simplifying assumption which could have been avoided by the use of random sampling. However, it was felt that the use of a random cycle would make the model only more complex, but not more useful.

orders arrive for the various product lines. Since most orders are obtained from customers who order the same product line repeatedly, there is bound to be some regularity in the production sequence. This ordered sequence is difficult to observe, but cannot be denied to exist.

The market operation generates two sets of orders for each product line. One set of orders, called the A list, has to be fabricated during the current schedule period. The other set of orders, called the B list, is not due until a future schedule period, presumably the next, but may be used to make up efficient schedule patterns.[3] Random sampling from empirically-based theoretical distributions then determines the number of orders for a product line, the number of items in an order, width of each item and length of each item.

A flow chart view of the model is presented in Figure 2-2. It shows the simulation process sequence, input and output. The main program which controls the simulation, including the various subroutines, is shown in Figure 2-3.

---

[3]The use of an A list and a B list is again a simplifying assumption. In actual practice the A list was real, but the B list was only partially real and partially "fabricated." By the term "fabricated" we mean that the scheduler would "fabricate" or generate orders that would make up better schedules. He would generate these orders by contacting regular customers for potential orders of the type he was looking for, or if no commitment could be obtained, he would obtain permission to produce "needed" orders for inventory. The term "needed" order here means needed to make up efficient schedules.

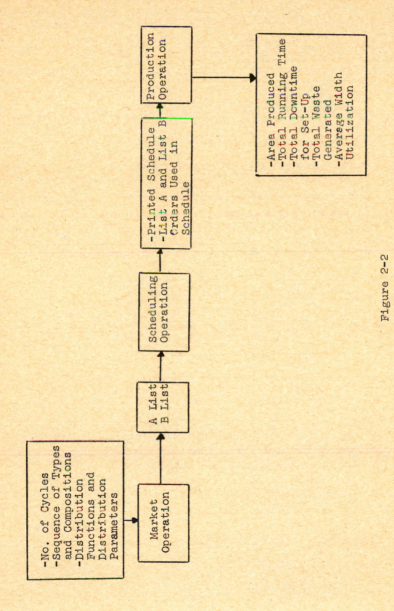

Figure 2-2

Flow Chart of Simulation

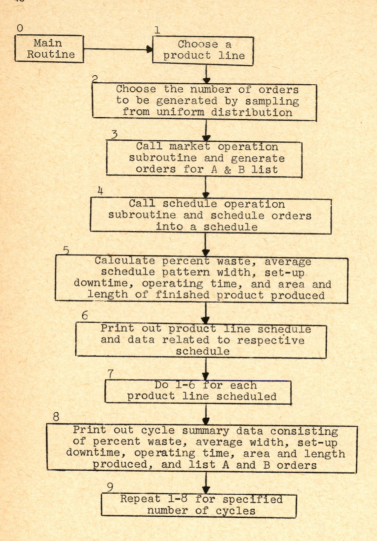

Figure 2-3

Flow Chart for Main Program

MARKET OPERATION

The market operation generates orders for a specific product line by random sampling. A multivariate, lognormal distribution is used because statistical tests indicated that the empirical data are lognormally distributed.[4]

Specification Parameters

The empirical order data suggested that some of the specification parameters might be related. Relations seemed to exist between the width and length of an item, between width and raw material composition, and possibly between the number of items per order and the above variables. As it was planned to generate the order data randomly from distributions based on empirical data, it would be important to

---

[4]The statistical tests used to check for normality were suggested by E.S. Pearson and H.O. Hartley, Biometrika Tables for Statisticians, Vol. 1, Cambridge: University Press, 1958, p. 61.

The tests are based on the functions of the moments of the observations. Two tests were used; the first test checks for skewness, and the second test checks for kurtosis. The 90% confidence limits and the values of the test statistics for the skewness test are respectively: number of items per order, $\pm.621$, $-.442$; length of item, ±459, $.345$; width of item, $\pm.459$, $-.447$. Only number of items per order and length of item satisfied the 90% confidence limits of the kurtosis test. The 99.0% confidence limits were used on width of item. The confidence limits and values of the test statistics are respectively: number of items per order, $.7440-.8578$, $.8334$; length of item, $.7583-.8403$, $.7902$; width of item, $.7313-.8629$, $.8623$.

take any relations between specification parameters into account.

To test the above hypotheses, the historical data were sampled to obtain 240 observations representing the twenty-five product lines, and a multiple regression routine was run on the sample data. The results of this regression are shown in Table 2-1.

The correlation between width of an item, length of an item and raw material composition is considered in the design of the market operation. This correlation can be handled by estimating distinct parameters for each product line or by calculating parameters as functions of raw material composition. Because it is difficult to calculate functional relations between the parameters and raw material composition, it was decided to estimate distinct parameters for each product line.

It has already been reported in the previous section how the product lines are sequenced during a cycle. The sequence was based on a detailed analysis which indicated that the high-volume product lines were generally scheduled three times as often as the low-volume product lines. Therefore, the model is instructed to follow a sequence which produces a cycle in which the nine high-volume product lines are scheduled three times and the sixteen low-volume product lines are scheduled once.

Table 2-1

Results of Correlation Test

Variables	95% Correlation Coefficient Interval	Uncorrelated
log (length) vs. composition	+.281 to .495	no
log (length) vs. log (items per order)	-.101 to .153	yes
log (length) vs. log (width)	+.184 to .416	no
log (items per order) vs. log (width)	-.227 to .024	yes
composition vs. log (items per order)	-.042 to .210	yes
composition vs. log (width)	+.142 to .379	no

Order Generator

To determine the number of orders that must be produced for each product line, the A list, random samples are
drawn from a uniform distribution. The distribution for each
product line is uniquely segmented so that the value of each
sample drawn is either zero or one. The number of samples
drawn is empirically-based. Summing the sample values then
determines the number of A list orders for a product line.
The B list has 10% fewer orders than the A list.

The next step is determining for each order the number of items ordered, the width of an item and the length of
an item. Number of items, q, is determined by drawing a
standard normal deviate, z, and using the formula derived by
Aitcheson and Brown.[5]

$$q = \exp\ (\mu_q + \sigma_q z),$$

where μ_q and σ_q are the parameters of the lognormal distribution for number of items. Similarly, width of an item, w,
is determined by the formula

$$w = \exp\ (\mu_w + \sigma_w z),$$

where μ_w and σ_w are the parameters of the lognormal

[5] J. Aitcheson and J.A.C. Brown, The Lognormal Distribution. Cambridge: University Press, 1957, p. 28. Note that
μ_q and σ_q are not the mean and standard deviation.

distribution for width. And finally, length of an item, v, is chosen on the basis of the width. In statistical terms v is conditional on w. The formula derived by Freund[6] is

$$v = \exp\left[\mu_v + \rho\,\frac{\sigma_v}{\sigma_w}\,(\log w - \mu_w) + \sigma_v\,z(1 - \rho^2)^{\frac{1}{2}} \right]$$

where μ_v, σ_v and ρ are the parameters of the lognormal distribution for length given width. A flow chart of the order generator is presented in Figure 2-4.

SCHEDULING AND PRODUCTION OPERATIONS

The scheduling operation develops schedule patterns consisting of one, two, or three orders. This operation is performed by a main scheduling subroutine with the aid of a second-level subroutine. The main subroutine schedules patterns consisting of one and two orders, and the second-level subroutine schedules the three-order patterns. Flow charts of the two scheduling subroutines are shown in Figures 2-5 and 2-6.

The orders for each product line are generated by the market operation and are presented to the scheduling subroutine in the form of an A and a B list. The A list must be scheduled during the current schedule period, that is, the A list orders have been promised for an early

[6] J.E. Freund, Mathematical Statistics. Englewood Cliffs, N.J.: Prentice-Hall, Inc., 1962, p. 305.

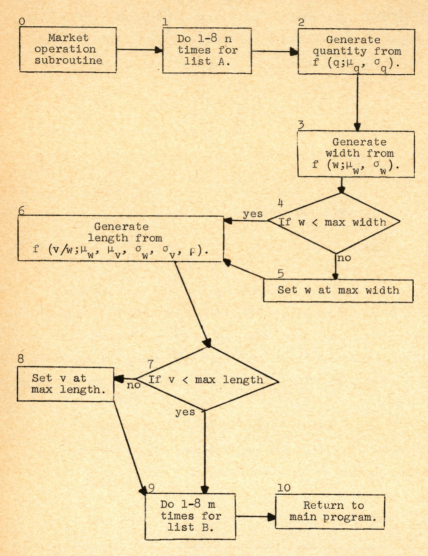

Figure 2-4
Quantity, Width, and Length Generation

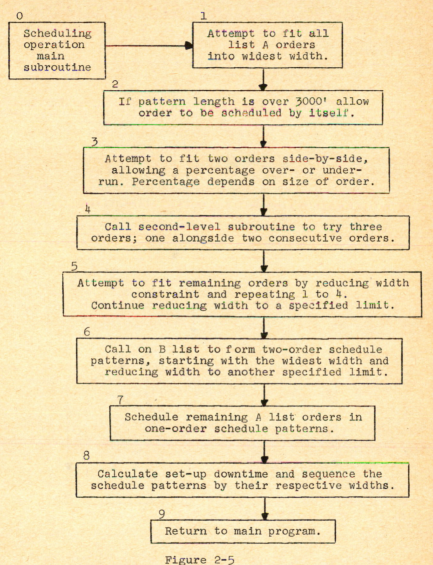

Figure 2-5

Scheduling Operation Main Subroutine

48

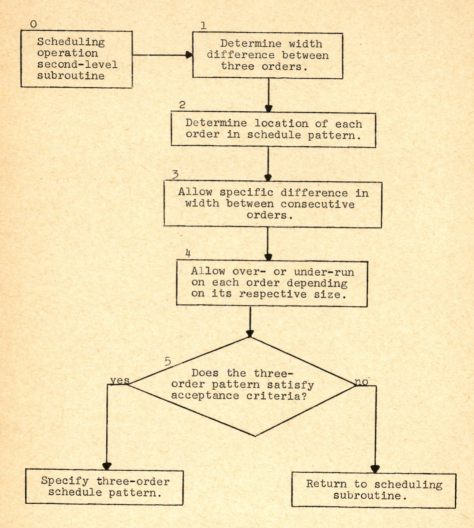

Figure 2-6

Scheduling Operation Second-Level Subroutine

delivery date; the B list on the other hand is a list of orders that may be scheduled during the current schedule period but is not due until a later date.

The B list is only consulted after it is found that an acceptable schedule pattern cannot be generated from the A list. At the start of the scheduling process for each product line, only A list orders are considered, and the imposed criteria for an acceptable schedule pattern are very tight. The acceptability criteria are relaxed, and as fewer A list orders are left to be scheduled, the B list is consulted for orders that will match the remaining A list orders.

Acceptance Criteria

The criteria that determine whether a schedule pattern is acceptable consist of a pattern length constraint, two width constraints, a width-match constraint, and an over- and under-run constraint. The purpose of the constraints is to obtain efficient schedule patterns and, at the same time, to maintain a rough analogy to the human scheduler of the empirical system.

Pattern length is determined by the length of the finished product strip required to fill an order. Hence, it is a function of number of items in the order, the width of an item and the length of an item. The pattern length constraint only applies to one-order schedule patterns, and the constraint is only enforced until it is found impossible

to schedule the remaining A list orders by themselves or
with B list orders.

The width constraints consist of an upper constraint
determined by the machine width capacity and a lower constraint intended to force high machine utilization. The
lower constraint starts out at 99% of the upper constraint
and is relaxed in 1% increments.

The width-match constraint only applies to three-
order schedule patterns. It requires that orders II and III
(in the three-order pattern in Figure 2-1) be nearly the
same width.

The over- and under-run constraints are determined
by a negative exponential function. The function allows
a small percentage over- or under-run on large orders and a
large percentage over- or under-run on small orders. The
maximum percentage, for very small orders, is about twenty-
five percent.

Production Operation Calculations

The production operation determines the values of the
output variables for each product line. These variables are:

    Total run length, in lineal feet.

    Total gross area produced, in square feet.

    Total waste incurred, in square feet.

    Percentage waste incurred.

    Total machine time, in minutes.

Total set-up downtime, in minutes.

Average width of scheduled patterns.

Number of A list orders used in the schedule.

Number of B list orders used in the schudule.

The above calculations are repeated for each schedule
period in the cycle. Upon completion of each cycle, the
following additional calculations are made for the cycle.

Total set-up downtime, in hours.

Total machine time, in hours.

Total length of finished product produced, in lineal
feet.

Total gross area of finished product produced, in
square feet.

Total waste incurred, in square feet.

Percentage waste incurred.

Average width of scheduled patterns.

Total number of A list orders.

Total number of B list orders.

VALIDATION

Developing a simulation model of a complex operation
is no end in itself. The model should be useful for testing
hypotheses and for investigating the effects of specific
changes.

Before these tests can be performed the model must
be shown to be a valid representation of the system being

simulated. The question of whether a model is valid is rather tenuous. All models are really approximations of the real world, and it is not possible to have an exact replication. However, the behaviors of certain variables should reasonably replicate the corresponding variables in the real world. This section analyzes these variables, and compares their properties in the simulation model to those in the empirical system.

Table 2-2 compares the seventeen variables. These seventeen variables were selected because they are expected to affect the operating and raw material cost output of the model if changes are made in the input variables for experimentation. All the variables selected are output variables. Input variables are not compared because the simulation input variables must reasonably correspond with the empirical input variables in order to obtain valid output variables.

No statistical tests were performed on the output data shown in Table 2-2. The objective of the validation was to obtain two comparable output variable vectors of the empirical process and the simulation model. In a perfect simulation where there is a one-to-one correspondence in input variables and internal relationships, one would expect two comparable output vectors. Statistical tests performed on the respective elements of these two vectors should indicate that each corresponding set of variables could have been drawn from the same distribution. However,

there are very few perfect simulations. Nature is complex
and defies to be perfectly imitated. Hence, simulations,
as a rule, are only approximations of the real process.
The reader will at this point realize that the simulation
of our complex industrial system is only an approximation;
many simplifying assumptions had to be introduced. Therefore,
the results in Table 2-2 may not be able to stand up to a
rigorous statistical test. However, they should be reasonably alike so that the model can be used for studying
the effects of various policy changes. It was decided that
the results shown in Table 2-2 satisfy the somewhat philo-
sophical statement of "being reasonably alike."

A complete listing of the computer program of the
simulation written in Fortran is listed in the general
appendix.

Table 2-2

Summary of Simulation Model Validation

	Empirical Process*	Simulation Model***
percentage waste incurred$	4.85%	4.97%
average machine width utilization as a percent of width capacity**	88.5%	88.3%
number of different raw material widths and compositions used	60	60
mean of schedule pattern length in feet	11450	10360
standard deviation of schedule pattern length in feet	14300	11200
number of raw material compositions and width changes per million square feet	8.0	8.7
number of changes from narrow to wide width per million square feet	1.25	1.16
number of type changes per million square feet	1.25	1.16
length of a cycle in millions of square feet	10.5	11.2
percentage of time machinery was down for set-up$$	15.0%	14.8%
machine hours per million square feet$$	12.6	12.9
number of orders per million square feet	24.7	23.4
percentage of type A produced	19	20
percentage of type B produced	35	33
percentage of type C produced	43	44
percentage of type D produced	3	3
percentage of medium weight of product produced	19	20

55

*Figures were estimated from large samples of
empirical data except where otherwise indicated.

$For empirical process based on estimate by
production manager.

**For empirical process sample estimate is identical
to estimate by production manager.

$$Based on accounting data.

***Figures were estimated from large samples of
simulation runs.

EXERCISES

1. Explain why in the market operation width and length, width and composition and length and composition are causally related.

2. Why is it important that the B list in the market operation be properly controlled?

3. If B list products were easily marketable, possibly at a discount, what changes in the scheduling operation would you expect to see or possibly recommend?

4. How would you simulate a job shop operation analogous to the one in this chapter but without a discernible cycle?

5. Why is validation of a simulation model so important, especially in the case of the model discussed in this chapter?

6. Which output variables in Table 2-2 could be compared and tested statistically assuming that sufficient historical data were available?

7. In chapter 1 simulation was defined as the building of an abstract model of a system and performing experiments on the model. Reflect on this definition and how does it apply to chapters 2 and 3 you just finished reading?

8. Corrugated Container Corporation has seven plants and it is planning to investigate if there are cost savings in combining any of the seven plants into larger and more efficient plants. Combining two or more plants,

however, will increase transportation costs considerably for the regional area being served by the plant which is being closed down. In addition, the plant which absorbs another plant will certainly need expansion. How would you develop a model to study and evaluate this situation for Corrugaged Container Corporation?

9. Suppose Containers, Inc. had two identical plants and markets spaced two hundred miles apart. The identical markets and plants are represented by the study described in chapters 2 and 3. How would you evaluate a proposal to close one of the plants and have both markets served by one plant located in either location? What information would you need to make this study?

CHAPTER 3.--ANALYSIS OF AN INDUSTRIAL SYSTEM WITH THE AID OF A SIMULATION MODEL[1]

Designing and building simulation models is a challenging task which can be quite rewarding from an achievement point of view. However, the purpose of **design and development of a simulation model** is usually to test out hypotheses or to perform some experiments. These hypotheses cannot be economically tested out on the real system, and therefore a simulation model is developed.

EXPERIMENTATION

This section will review the various uses of a validated simulation model. The model used is the one developed in the previous chapter. The range of experiments performed is by no means exhaustive, but does cover a wide variety of experiments. A total of six of these experiments are described below.

Policy of Minimum Order Size

At present the firm accepts orders of any size. What would be the consequence of imposing a minimum order size of 10,000 square feet? Such a policy should have effects on

---

[1]This chapter is based on the second part of C. Carl Pegels: "Simulation and the Optimal Design of a Production Process," The International Journal of Production Research, Vol. 7, No. 3 (1969), pp. 219-231.

four characteristics of the manufacturing process: percent
waste, machine utilization, set-up downtime and machine
hours.

The minimum order size was imposed on the model and
a simulation run provided the following answers: there was
no change in percentage waste, no change in machine width
utilization, a 26% reduction in set-up downtime from 14.8%
to 10.9%, and a 4% reduction in machine hours from 12.9 to
12.4 hours per million square feet.

Will the minimum order policy affect market demand?
The model can supply information on what percentage of orders are under 10,000 square feet. However, the marketing
department would have to estimate how the minimum order
size policy would affect demand. Some customers with small and
larger orders would probably switch to other suppliers. Since management
knows the savings, it can compare the expected drop in profits, resulting
from dissatisfied customers if the policy is adopted, and thus
make decisions on the basis of more and better information
than without the aid of the simulation model.

Changes in Set-Up Times

Change-over set-up allowances[2] for each order--or,
if two orders are completed simultaneously, for each set of

---

[2]Change-over allowances are considered to be identical
to actual times required to perform the set-up in this experiment.

two orders--are functions of the technological level of the
machinery. This experiment determined the effect of changing these set-up allowances by making technological changes.
A simulation model is especially useful for this experiment
since part of the set-up work is performed while the machine
is running.

The set-up allowances were changed in 12.5% increments
with the 100% level representing the present allowances. The
dependent variables were set-up downtime and machine hours.
The results of this experiment are presented in Table 3-1.

Table 3-1

Results of Change-Over Time Allowances

Change-Over Allowance (Percent of Validation Run)	Percent Downtime For Set-Up	Machine Hours (per Million Square Feet)
37.5	12.7	12.6
50	14.0	12.7
62.5	13.7	12.1
75	13.6	12.5
87.5	14.5	13.0
100	14.8	12.9
112.5	17.4	12.9
125	16.1	13.0
137.5	16.8	13.1
150	18.7	13.2

Varying Operating Speeds

Increasing the machine operating speeds of the existing production facility or acquiring a new production facility with higher speed capability poses the important
question: Will the savings justify the required capital
outlay?

The effects of nine different operating speeds on
set-up downtime and on machine operating hours are presented
in Table 3-2. The speeds were varied in 10% increments with
the 100% level representing the present speeds.

Table 3-2

Results of Machine Operating Speed Changes

Machine Operating Speeds (Percent of Validation Run)	Percent Downtime For Set-up	Machine Hours (per Million Square Feet)
80	11.8	15.4
90	13.8	14.1
100	14.8	12.9
110	17.5	11.9
120	19.3	11.5
130	22.7	11.0
140	25.6	10.3
150	24.6	9.7
160	27.1	9.3

Varying the Lengths of Production Runs

At present, the simulation plant schedules each product line at least once during a cycle and nine of the twenty-five product lines are scheduled three times during a cycle. What would the effect be on the output characteristics if this policy were changed to schedule each product line only once during a cycle?

An experiment was run assuming that the number of orders for each product line would remain the same as at present. This experiment resulted in a 12% reduction in waste from 4.97% to 4.37%, in a 2% increase in machine utilization from 88% to 90%, in a 29% reduction of set-up downtime from 14.8% to 10.5%, and in a 6% reduction in machine hours from 12.9 to 12.1 hours per million square feet.

The implications of this experiment can be weighed only in the context of information about the market effects of such a change. The average order delay over and above normal delays is less than two working days, and the maximum order delay is less than 3.5 working days. With the above information the sales department could estimate the effects on demand of slower deliveries.

Increasing the Machinery Width Capacity

Suppose the firm wants to estimate the benefits of buying machinery with a higher width capacity. To obtain the required estimates, simulation runs with various width capacities were made. The machine width capacity was varied from 100%, the present width, to 190% in 15% increments. The changes in the machine width capacities affected five characteristics of the process: waste, machine utilization, set-up downtime, machine hours and percent list B orders. The results are shown in Table 3-3.

Table 3-3

Results of Machine Width Capacity Changes

Machine Width (Percent of Validation Run)	100	115	130	145	160	175	190
Percent Waste	4.97	4.21	3.99	4.00	3.67	3.72	3.51
Percent Machine Width Utilization	88.1	89.5	90.1	89.8	87.8	89.5	91.4
Percent Downtime for Set-up	14.8	19.2	21.8	27.2	29.5	36.8	42.3
Machine Hours (per Million Square Feet)	12.9	11.7	10.4	10.2	10.2	9.9	9.4
Percent List B orders	13.0	14.2	14.4	10.9	9.8	6.8	5.6

A further analysis is required of the costs and benefits of different machine widths to determine the optimal machine. A wider machine provides savings in waste and in machine hours per million square feet. However, percent down-time for set up increases markedly and the additional capital cost of the new machine must of course also be considered. The above experiment assumes that the market has remained unchanged. However, with a higher machine width capacity, orders for a wider product may be accepted and hence, machinery with a wider width capacity may also have significant marketing benefits. If the additional marketing benefits are known before hand or if they can be estimated then the simulation model can be used to determine the savings in waste reduction and in machine running time.

DETERMINING OPTIMUM MACHINE CHARACTERISTICS

A method will be discussed which combines the simulation approach and the analytic approach to determine optimum capacity and performance character-istics for a machine to process orders for this market. Although a large number of simulation runs with different machine characteristics could also be used to search for the optimum characteristics of the production facility, the use of simulation by itself has a serious disadvantage. This disadvantage is that the outputs of each individual simulation run are randomly distributed. Runs must be repeated several times before determining in which direction to adjust the input variables. The combined approach works with an expected total cost function which facilitates the search for the optimum point.

In this example, simulation is used to generate output data on a number of machine characteristics. The output data are then summarized in equations obtained by multiple regression techniques.[3] These equations predict waste as a function of machine width and predict machine hours as a function of machine width and speed. The predicted waste and machine hours are then multiplied by their respective cost functions to estimate waste cost and machine operating cost per unit quantity of finished product. The various costs are added, and optimization methods are applied to obtain the minimum cost values of machine width and speed.

Cost and Functional Relationships

The total cost function is broken down into three components. The first of these, expected waste cost, is formulated as follows:

$$E(WC) = k_2 f_1(W).$$

k_2 is raw material cost per unit area. W is the machine width. $f_1(W)$ is the predicted waste as a quantity of raw material per unit quantity of finished product. Its functional form is postulated to be:

$$f_1(W) = a_{10} + a_{11}W + a_{12}W^2.$$

The second cost component, expected operating cost, may be expressed as:

$$E(OC) = g_1(W,S) \, f_2(W,S).$$

S is the machine operating speed, and $g_1(W,S)$ is the operating cost per machine hour. It is estimated from machinery

[3]A similar approach for estimating production functions was used by O. Heady, "An Econometric Investigation of the Technology of Agricultural Production Functions," Econometrica, Vol. 25, April 1957.

manufacturers' data. Its functional form is assumed to be:

$$g_1(W,S) = b_{10} + b_{11}W + b_{12}S.$$

The function $f_2(W,S)$ is the predicted operating hours per unit quantity of finished product. Its functional form is postulated to be:

$$f_2(W,S) = a_{20} + a_{21}W + a_{22}S + a_{23}W^2 + a_{24}S^2 + a_{25}WS.$$

The third cost component, expected capital cost, may be expressed as:

$$E(CC) = g_2(W,S).$$

$g_2(W,S)$ is the capital cost per unit quantity of finished product. This is a deviation from the usual practice of expressing capital cost per unit time. However, if annual production figures are known, one function can be derived from the other. The functional form is assumed to be:

$$g_2(W,S) = b_{20} + b_{21}W + b_{22}S + b_{23}W^2 + b_{24}S^2$$

Total Cost Function and Optimization

Adding the three cost components provides the expected total cost function:

$$E(TC) = E(WC) + E(OC) + E(CC).$$

The expected total cost function may be minimized by taking partial derivatives with respect to W and S. Setting these partial derivatives equal to zero provides two simultaneous equations in two unknowns. Due to the complexity of these equations they cannot be solved analytically. However, gradient methods provide a very rapid solution technique.

An example, based on the model, was designed and programmed on a computer. The validation run of the simulation model was used as the basis of this example. Cost data and cost functions, which are in agreement with real operations, have been used. The expected total cost function has the form:[4]

$$E(TC) = c_{10} + c_{11}W + c_{12}S + c_{13}WS + c_{14}W^2 + c_{15}S^2 +$$
$$c_{16}W^2S + c_{17}WS^2 + c_{18}W^3 + c_{19}S^3$$

subject to: $100 \leqq W \leqq 190$

and $80 \leqq S \leqq 160$

The gradient solution technique revealed that the optimum width is 155.0% of the present width and the

---

[4] The values of the coefficients are: $c_{10} = 1209.5$, $c_{11} = 1.808$, $c_{12} = 2.16$, $c_{13} = .02589$, $c_{14} = .09699$, $c_{15} = .00379$, $c_{16} = .0000648$, $c_{17} = .0000693$, $c_{18} = .0000508$ and $c_{19} = .0000248$.

optimum operating speed is 121.4% of the present run speed.
On the basis of the assumed cost functions, the plant would
be more efficient if the present machinery were replaced
by machinery with a larger width capacity and a higher
operating speed.

CONCLUSIONS

A simulation model of a complex industrial system has
been used to experiment with various potential changes in
the production system which was too complex to be studied
solely by analytic techniques. Several illustrative experimental applications of the simulation model were presented, including a model which combines simulation and
analytic techniques to find the optimum width capacity and
speed of a production facility.

Detailed information on any of the examples would not
be available without a simulation model of the actual system.
Although estimates could have been made, these would be subject to considerable error because of the complexity of the
system.

The changes that were selected and analyzed only
provide a small sample of numerous changes that can be
analyzed by the simulation model. The model will be especially useful for analyzing the effects of proposed capital investments, which will alter the system's operating
characteristics.

EXERCISES

1. The evaluation of the minimum order size policy resulted in a reduction of 26% in set-up downtime and in a 4% reduction in machine
   hours from 12.9 to 12.4 hours per million square feet. If the cost of
   running the machine is $200 per hour, how much of a surcharge should
   be imposed on orders under 10,000 square feet to pay for the added
   cost of operating the machine with the small orders?

2. Is a surcharge advisable in exercise 1 or what other alternatives do
   you propose?

3. Would you advise the expenditure of funds to reduce the set-up allowances on the basis of the simulation results in Table 3-1? Elaborate.

4. If the cost of operating the present machine is $200 per hour, what
   maximum amount per hour should the firm be willing to pay for a
   machine that is 40% faster? Base your analysis on the results in
   Table 3-2.

5. If the plant operates 80 hours per week what would be the weekly
   savings of running the plant if each product line was only run once
   during a cycle?

6. On the basis of an 85 hour week for the present machine, what would
   be the savings of running a machine with a width capacity 30%
   larger than the current machine?

Assume that additional orders can be obtained to run the wider machine.
Operation cost of the current machine is $200 per hour and $220 per hour
for the wider machine. Waste cost is $0.01 per square foot.

7. Computer Control is planning to launch a time sharing computer facility in
a new regional market. It is presently in the time sharing business in
another regional area and has gained considerable experience with the
marketing of computer services. The new computer facility has considerably
larger capacity than their current one and therefore a considerably larger
number of customers for the new facility seems warranted. At the same
time Computer Control does not want to overload its computer facility
causing customer delays and possible customer losses. Hence it has
decided to run a simulation to determine customer delays at critical heavy
load periods. Time sharing demand is heaviest during afternoon hours,
lightest at night and average in the morning. A typical customer will use
the computer on a workday during the morning, afternoon and evening hours
with the following probabilities.

Time Period	Probability
8 to 10 am	.03
10 to 12 noon	.10
12 to 2 pm	.20
2 to 4 pm	.30
4 to 6 pm	.25
6 to 8 pm	.05
8 to 10 pm	.05
10 to 12 pm	.02

The time sharing computer can accommodate up to 35 customers at a time.
Any customers attempting to connect while the computer is fully connected

are told that they must wait. The amount of computer central processor time used also varies considerably. Some customers use little whereas others use a large amount. There is no real relationship between light and heavy users and time of day. However, the probability of a light, medium or heavy user is known and is respectively 0.40, 0.35, and 0.25. If in excess of 15 heavy customers or 20 medium customers are connected, serious delays are incurred by customers connected to the computer. These delays tend to turn customers away and therefore should be avoided or minimized. Based on the above scenario develop a computer model (simulation) to study the proposed time sharing computer installation of Computer Control.

8. Now that you have designed a computer model (simulation) of Computer Control's time sharing computer problem in exercise 7, how does it compare with the complex model discussed in this chapter? Is your model comprised of a market operation, a scheduling operation and a production operation? What advantages are there in modular design as suggested above especially for the time sharing computer operation?

9. Ace Tool and Die is considering the acquisition of several pieces of numerically controlled machinery to replace manually operated equipment. The work load of the plant is fairly predictive on the basis of previous year's operations. The only variation from year to year is usually in utilization of machinery. That is, bottlenecks occur every year but not

always at the same work center. The plant has a total of sixteen work centers and four of the work centers which have been frequent bottlenecks in the past are candidates for the new machinery. The new machinery has twice the capacity of the machinery it replaces and is also considerably cheaper to operate. The main objective of Ace Tool and Die, however, is the reduction in bottlenecks so that costly delays can be reduced. Design a simulation to study Ace's problem. Would your simulation be modular? What kind of information would you need to operate the model?

10. Using the information of Ace Tool and Die in problem 9, develop a validation procedure to test out the model you designed.

11. The Ace Tool and Die simulation (see exercises 9 and 10) really consists of several stages. What are these stages? Hint: Refer to chapter 1 for simulation definition.

12. The simulation discussed in this chapter uses the modular approach. That is it has broken down the total process or system in several subsystems such as the production process, the scheduling process and the market process. Why was this done? What are the advantages of modularization?

13. What other products could use the simulation model, or a modified version, described in this chapter?

14. Would you want to validate a computer model (simulation) of a coin tossing experiment? of a die rolling experiment? Discuss and state your reasons.

CHAPTER 4.--ANALYSIS OF A PLANT ELECTRICAL POWER SYSTEM[1]

In the previous three chapters the topic of systems analysis was introduced and a simulation model of a complex industrial system was developed and validated. Subsequently the simulation model was used to perform several experiments. In this chapter discussion on the topic of systems analysis will be continued, although a different system will be presented and different methods of analysis will be used. The system to be discussed in this chapter is still in an industrial environment, but is not directly related to production as the previous system was. The system to be analyzed in this chapter is a subsystem of an industrial plant's electrical power system. The problem to be resolved is how to find the minimum cost level of reactive power consumption.

THE PROBLEM

Engineering improvements in the electrical systems of industrial plants can decrease electric power costs significantly. These improvements consist of using capacitors to reduce the demand for reactive power. However, capacitor installations are costly and the question arises to what extent the power factor should be corrected. In addition,

---

[1] This chapter is based on C. Carl Pegels: "A Management Science Approach to Power Factor Correction," The International Journal of Electrical Engineering Education, Vol. 5 (1967), pp. 691-97. (published by Pergamon Press Ltd.)

demand for reactive power usually varies from month to month, or from season to season.

Use of Capacitors to Reduce Reactive Power Demand

Reactive power is the electricity used in magnetizing the iron in electrical machinery. This reactive power does not contribute to useful work output, because capacitors can supply this reactive power. Although reactive power does not contribute to useful work, it is costly for the electric power utility supplying the power. Hence, penalty charges are levied by power utilities if the reactive demand exceeds a specified fraction of normal demand, or charges for reactive demand are based on a sliding scale, i.e., low charges per KVAR[2] for minimal demand, and higher charges for higher demands. Capacitor installations also increase the current carrying capacity of existing wiring, transformers, and electrical switchgear. However, this benefit can only be realized if the capacitor installation is at or near the machinery using the reactive power. This is seldom feasible, and therefore this benefit has not been considered in the following analysis. This benefit does point out the reason why electric power utilities impose penalty charges for reactive power. Excessive reactive demand decreases the

---

[2]KVAR is the term for kilo volt amperes reactive. It is the unit measurement for reactive power demand.

current carrying capacity of the power utility's electric
power "transportation" (i.e. power transmission) facilities.

Location of Capacitors

Although locating the capacitor at the electrical
machinery has already been assumed infeasible, there still
remain two other options. These are: locate the capacitors
at the point where the electric power enters the plant, or
locate the capacitors at the load centers or at the intermediate transformers (load center).

To resolve the question of where to locate the capacitor or capacitors the following incremental economic
analysis is suggested. If the estimated savings realized by
locating the capacitors at the load center exceed the extra
cost of installing the capacitor at the load center then
make the capacitor installation at the load center.

THE MODEL

The main problem we are   concerned with is
the determination of the number of capacitors to install
so that overall cost to the firm is minimized. Another way
to explain the problem is to ask up to how many capacitors
should be installed so that the minimum rate of return on
investment (in capacitors and installation) is satisfied.
In other words, the industrial firm utilizing the proposed
model must, among others, specify its required minimum rate

of return for new investments. It will be assumed that this minimum rate of return is also its cost of capital.[3] This assumption must be made, otherwise the model can not claim to minimize overall cost.

The proposed model will utilize historical reactive power demand data and/or projected subjective data to calculate the present value of expected penalty charges for reactive demand. The number of capacitors to be installed will then be determined by minimizing the expected total cost of the capacitors, the capacitor installations and the penalty charges. The capacitors that are installed will then provide a rate of return equal to the interest rate used to calculate present value. Figure 4-1 presents a graphical picture of this decision making model.

Cost Curves

The cost curves in Figure 4-1 are derived as follows. If no capacitors are installed, the cost of the capacitors and installation is nil, and the present value[4] of future

---

[3]The cost of capital or the rate of return is used to calculate the present value of the stream of reactive demand charges that will be incurred if no capacitor installations are made.

[4]The appendix of chapter 7 discusses in extensive detail how the present value equivalent of a cash flow is found. The reader may also want to refer to R.K. Davidson, V.L. Smith and J.W. Wiley: Economics: An Analytical Approach, Homewood, Illinois: Richard D. Irwin, Inc., Revised Edition, 1962, p. 116.

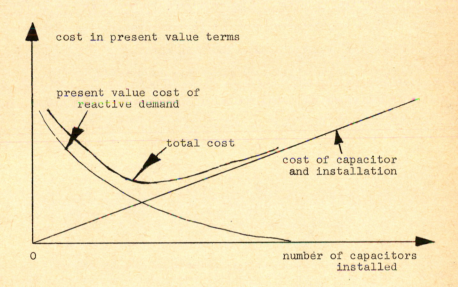

Figure 4-1

Cost Curves for Reactive Demand and for
Capacitors and Installation Charges

penalty charges for reactive power demand, C_R, is:

$$C_R = \sum_{i=1}^{T} \frac{x_i}{(1+r)^i}$$

where x_i = penalty charges per month

r = interest rate or cost of capital per month

T = life of capacitors in months

Since interest is compounded very frequently, monthly, the present value for penalty charges for reactive power demand can be approximated by the following expressions:

$$C_R = \frac{1-\exp(-rT)}{rT} \sum_{i=1}^{T} x_i$$

$$= \frac{\bar{x}[1-\exp(-rT)]}{r} \qquad (4-1)$$

where $\bar{x} = \frac{1}{T} \sum_{i=1}^{T} x_i$ is an estimate of the average monthly penalty charge.

If the life of a capacitor is expected to last for the life of the plant, or the life of the machinery it serves, then a reasonable simplifying assumption is to set T equal to infinity. This will simplify the cost function (4-1) to:

$$C_R = \frac{\bar{x}}{r} \qquad (4-2)$$

Now that a rather simple cost function has been derived, it is not very cumbersome to calculate the present value of reactive power demand penalty charges for each incremental capacitor. The present value costs are plotted and the cost curve shown in Figure 4-1 is derived. Similarly, the costs of each incremental capacitor and installation are plotted. Adding the cost values together for each incremental capacitor installation provides a total cost curve with its minimum cost value as shown.

In the above analysis it is assumed that the reactive power charges are a deterministic value. This is not realistic; the penalty charge per month is a random variable. In addition the incremental cost increases rapidly as the reactive power demand increases. Hence, it is possible to obtain an average penalty charge per month, but it is not easy to determine the average reactive demand per month. Under these circumstances an alternative approach is desired.

An Alternative Approach

One proposed approach estimates the distribution of the random variable, reactive demand per month, from historical demand data. In addition the cumulative present value of penalty charges are used to estimate a cumulative cost function. If the estimated distribution and estimated cumulative cost function are compatible, it is relatively straightforward to determine the expected cumulative penalty charges for any number of capacitors installed.

Suppose the estimated distribution of reactive demand is f(x), and the cumulative cost function is g(x), then the expected value of the monthly reactive demand charge is:

$$E[g(x)] = \int_{-\infty}^{\infty} g(x)f(x)dx \qquad (4-3)$$

Compatibility of these two functions should not pose a serious problem since the density function f(x) is most likely normal, uniform or exponential.[5] The cost function is expected to be exponential and (4-3) can be applied quite readily (see appendix) to calculate E[g(x)].

AN EXAMPLE

Suppose that analysis of historical records reveals that the average reactive demand per month is 1800 KVAR.

---

[5] One technique for estimating the density function for the random variable, reactive demand, is to group the historical reactive demand data in equal intervals, and then plot the frequency with which an observation falls in each interval. From the derived graph a density function can be estimated. If sufficient data is available a statistical technique, called the chi-square test, can then be applied to determine if the data statistically fits the estimated density function. This test is described in any good textbook on statistics. If the data is limited and if it appears that the density function is normal then a better test is the one suggested and described by E.S. Pearson and H.O. Hartley, Biometrika Tables for Statisticians, Vol. 1, Cambridge: University Press (1958), p. 61.

The monthly penalty charges are determined by the rate schedule shown in Table 4-1. This rate schedule in turn is based on the firm's demand for active power. Hence, it is not typical, but it is a function of active demand.

Table 4-1

Penalty Charges for Reactive Demand

KVAR	Penalty Charge per KVAR	Cumulative Cost at Upper Limit
0-700	$.01	$ 7
701-1000	.02	13
1001-1200	.03	19
1201-1300	.05	24
1301-1400	.08	32
1401-1500	.12	44
1501-1600	.16	60
1601-1700	.20	80
1701-1800	.24	104
1801-1900	.28	132
1901-2000	.32	164
2001-2100	.36	200

If an interest rate, or cost of capital, of one percent per month is assumed, and the facility is assumed to have an infinite life, then using cost function (4-2) the present value of penalty charges can be calculated.

Assume that the capacitors to be installed are rated at 20 KVAR. Hence, for each capacitor installed in the plant's electrical system, the reactive demand charge will

be lowered by 20 KVAR per month. To eliminate all reactive
demand charges will therefore require the installation of
at least 90 capacitors. However, the reactive power is only
for the demand during the month and if in any monthly period
reactive demand is less than required, credit is not received for having excessive reactive demand capacity. Explained in another way 90 capacitors would only guarantee
that no reactive demand charge is incurred 50% of the time.

Regardless of whether a charge is incurred for reactive demand in any month, the main concern of the model is
to install that number of capacitors that will provide minimum overall cost. Assume that price discounts are available
if capacitors are acquired in volume. Similarly installation
costs vary with the number of capacitors installed. The
result is a listing of capacitor installation charges as
depicted in Table 4-2.

Using the data in Table 4-1 one can estimate the cost
function, g(x), where x is the variable KVAR's demanded.
One algebraic function that is expected to fit the present
value of cumulative cost data well is the positive exponential function,

$$g(x) = a \exp(bx)$$

To finalize this example, assume that the reactive
demand per month is normally distributed[6] with mean μ and

---

[6]The reader not familiar with probability distribution and probability
density functions should refer to any introductory Statistics textbook or to
J.E. Freund, Mathematical Statistics, Englewood Cliffs, N.J.: Prentice-Hall,
Inc., 1962.

Table 4-2

Capacitor Installation Charges

Number of Capacitors	Capacitor Cost		Installation Cost		Cumulative* Total Cost
	Increm.	Cumul.*	Increm.	Cumul.*	
1	$200	$ 200	$100	$ 100	$ 300
2	200	400	50	150	550
3	200	600	40	190	790
4	200	800	30	220	1020
5	200	1000	20	240	1240
6-10	160	1800	20	340	2140
11-20	160	3400	20	540	3940
21-30	160	5000	20	740	5740
31-40	150	6500	20	940	7440
41-50	150	8000	20	1140	9140
51-60	150	9500	20	1340	10840
61-70	150	11000	20	1540	12540
71-80	150	12500	20	1740	14240
81-90	150	14000	20	1940	15940
91-100	150	15500	20	2140	17640

*All cumulative figures are calculated at upper limit for intervals.

standard deviation σ. Then,

$$f(x) = (2\Pi)^{-\frac{1}{2}} \sigma^{-1} \exp\left[-\frac{1}{2}\left(\frac{x-\mu}{\sigma}\right)^2\right]$$

The expected value of the monthly demand charge can now be readily determined using equation (4-3).

The critical reader at this point may have observed that the estimation of density and cost functions could be skipped because a detailed listing of reactive demand

charges is presented in Table 4-1. This is true, but an
organized methodical approach as suggested above is intuitively more appealing, and also enables the researcher
to quickly determine what will happen if certain exogenous
or endogenous conditions are altered.

Plotting the cost data for various quantities of
capacitors generates the two cost curves as shown in Figure
4-2. From observation of the total cost curve the minimum
cost number of capacitors can be derived. An easier itera-
tive procedure is to calculate the total cost for each incremental capacitor or set of four or five capacitors,
starting with zero capacitors the penalty charges and the
capacitor and installation charges are calculated; then
this procedure is repeated until the minimum cost number of
capacitors is found. Table 4-3 utilizes this approach.

AN ALTERNATIVE METHOD

An alternate method is to estimate the algebraic forms
of the two cost functions in Figure 4-2. Suppose the penalty
charges for reactive demand, C_1, can be approximated by:

$$C_1 = \alpha \exp(-\beta x)$$

and the capacitor and installation cost, C_2, by:

$$C_2 = \gamma x$$

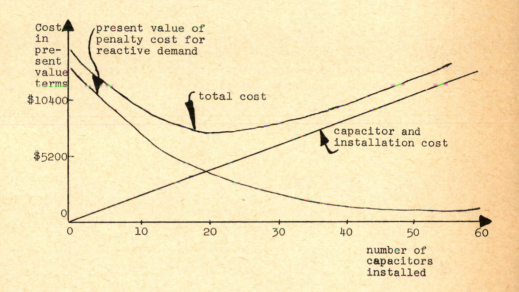

Figure 4-2

Cost Curves Based on Example

Table 4-3

Minimum Cost Capacitor Installation

Number of Capacitors	Present Value of Penalty Charge for Reactive Power	Capacitor Installation Charge	Total Cost
0	$10400	$ 0	$10400
5	8000	1240	9240
10	6000	2140	8140
15	4400	3040	7440
16	4160	3220	7380
17	3920	3400	7320
18	3680	3580	7260
19	3440	3760	7200
20*	3200*	3940*	7140*
21	3040	4120	7160
22	2880	4300	7180
23	2720	4480	7200
24	2560	4660	7220
25	2400	4840	7240

*Minimum cost number of capacitors.

where x is number of capacitors, and α, β, γ are parameters. Then total cost, C, is

$$C = C_1 + C_2 = \alpha \exp(-\beta x) + \gamma x$$

Taking the first derivative with respect to x,

$$\frac{dC}{dx} = \alpha\beta \exp(-\beta x) + \gamma$$

and setting the first derivative equal to zero, provides

the minimum cost solution

$$x = \frac{\log \alpha\beta - \beta \log \gamma}{\beta}$$

EXERCISES

1. Assuming the cumulative cost data for capacitor installation in Table 4-2 can be approximated by the positive
   exponential function $g(x) = a \exp(bx)$, find the parameters a and b by regression methods. Note that the
   exponential function can be written in logarithmic form
   as $\log g(x) = \log a + bx$.

2. Using the alternative method in the previous section and
   the data in Table 4-3 approximate the present value of
   penalty charges for reactive demand by the negative exponential function and the capacitor and installation
   charges by a linear function. Calculate the parameter
   values of α, β and γ by regression methods and find the
   minimum cost value of x. How does the value compare
   with the optimal value in the Table?

3. Assuming that the estimated number of capacitors to be
   installed is normally distributed with mean of 55 and
   standard deviation of 5, find the expected cost of capacitor installation. Use the positive exponential function estimated in exercise 1 and combine it with the
   normal density function.

4. Suppose that the present value of penalty charges for reactive demand are represented by the negative exponential function $f(x) = \alpha \exp(-\beta x)$ where $\alpha = 10000$ and $\beta = 0.06$. Capacitor installation charges are similarly represented by the positive exponential function $g(x) = a \exp(bx)$ where $a = 1$ and $b = 0.08$. Total cost can now be represented by $h(x) = \alpha \exp(-\beta x) + a \exp(bx)$. Find minimum cost value of x (number of capacitors to be installed).

5. Synthetic Textiles is averaging reactive demand charges of 2950 KVAR's per month. The penalty charge for the first 1000 KVAR's or less is $10 per month. For each incremental 200 KVAR's the penalty charge increases by $0.03 per KVAR. For example, the charge for 1200 KVAR's is $16, for 1400 KVAR's the charge is $28, for 1600 KVAR's the charge is $46, and so forth. To reduce the reactive demand charges capacitors may be installed. Each capacitor has the ability to lower reactive demand by 25 KVAR's. The installed cost of capacitors is the same as the listing in Table 4-2. What is the optimum (minimum cost) number of capacitors that should be installed?

6. If reactive demand charges for Synthetic Textiles (exercise 5) are normally distributed with mean of 2950 and standard deviation of 275, how many capacitors should be installed to ensure that montly reactive demand charges are only incurred with a probability of 1/3 ? How does the cost of this policy compare with the minimum cost policy in exercise 5?

7. If reactive demand of Synthetic Textiles averages 2950 KVAR's per month and is uniformly distributed with a range from 2300 to 3600 KVAR's per month find the number of capacitors required to ensure that reactive demand charges are only incurred once every six months. What is the cost of this policy?

8. Estimate the parameters for the negative exponential function, $C_1 = \alpha \exp(-\beta x)$, for reactive demand and the parameters of the capacitor and installation cost function of the form $C_2 = \gamma_1 x - \gamma_2 x^2$. Find the optimum (minimum cost) value of x.

9. Suppose that the power company supplying power to the plant in the example in the chapter gave a rebate to firms whose reactive demand does not exceed a specified figure each month. Based on the cost data given how large should this rebate be to be attractive to the power customers? Assume that reactive demand is normally distributed with mean of 1800 KVAR's per month and standard deviation of 200 KVAR's. Note this is a decision problem for the power supplier whereas the other exercises were decision problems for the power user.

10. Discuss in what other areas the models in this chapter could be used in the same or modified forms.

APPENDIX

Several density functions will be integrated jointly with the positive exponential cost function to derive the expected value $E[g(x)]$. Three cases will be considered.

Case 1 - Normal distribution:

Let $f(x) = (2\Pi)^{-\frac{1}{2}} \sigma^{-1} \exp\left[-\frac{1}{2}\left(\frac{x-\mu}{\sigma}\right)^2\right]$ $-\infty \leq x \leq \infty$

and

$g(x) = a \exp(bx)$, then

$$E[g(x)] = \int_{-\infty}^{\infty} g(x)f(x)dx = a \exp\left[\frac{b^2\sigma^2}{2} + \mu b\right]$$

Case 2 - Uniform distribution:

Let $f(x) = \frac{1}{\beta - \alpha}$ $\alpha \leq x \leq \beta$

and

$g(x) = a \exp(bx)$, then

$$E[g(x)] = \frac{a}{b(\beta - \alpha)} \exp(bx)$$

Case 3 - Negative exponential distribution:

Let $f(x) = \alpha \exp(-\beta x)$ $x \geq 0$

and

$g(x) = a \exp(bx)$, then

$$E[g(x)] = \frac{a\alpha}{b - \beta} \exp(b - \beta)$$

CHAPTER 5.--CAPITAL INVESTMENT ANALYSIS BY DECISION TREES

Decision tree analysis[1] has proven valuable in a variety of decision situations where a limited number of alternatives are explored and where additional information acquisition is one of the possible alternatives. It also forces an analysis of the total problem thus frequently producing additional alternatives which can be evaluated.

Investment analysis decisions usually involve large sums of money and the costs associated with a wrong decision could be sufficiently large to bankrupt some firms. It is therefore of utmost importance that the best decisions are made on the basis of the information available.

To be most effective several functional groups should be involved in an investment analysis. Marketing, finance, industrial engineering, product engineering and manufacturing should all be involved if all aspects of investment alternatives are to be explored. If each member of the investment analysis group is encouraged to question the estimates of other groups, the final completed analysis will prove to be

---

[1]Decision tree analysis has been explored by a number of authors such as Howard Raiffa: Decision Analysis, Reading, Massachusetts: Addison-Wesley, 1968; J.F. Magee: "How to Use Decision Trees in Capital Investment," Harvard Business Review, September-October 1964, pp. 79-96; and F.M. Bass: "Marketing Research Expenditures-A Decision Model," The Journal of Business, Vol. 26 (January 1963), pp. 77-90.

sound and decisions based on it will provide maximum expected profit.

The three examples which form the major part of this
chapter will reveal why a graphical decision tree analysis
is necessary if all possible alternatives are to be explored.
Although it is technically feasible to perform the analysis
in a tabular way, the graphical analysis of the decision
tree clarifies and follows the paths of all investment alternatives.

OIL WELL DRILLER PROBLEM

Before proceeding to an investment analysis a brief
illustration of a decision tree application will be provided. Suppose an oil well driller must decide whether to
drill a well at a given promising location or not. He has
assigned subjective probabilities to the well being a hit
or a miss of 0.40 and 0.60 respectively. He also has two
additional alternatives. Both alternatives are geophysical
surveys but of different levels of intensity and also of
cost. The one survey costs $50,000 and the other costs
$100,000. The first survey has the following historical
record; if there is oil there is a conditional probability
of 0.70 that the survey will so indicate. If there is no
oil there is a conditional probability of 0.40 that the
survey will indicate that there is oil. The second survey
has respective conditional probabilities of 0.80 and 0.30.

Table 5-1 lists the known probabilities of the two surveys.
The probabilities in the left column are simple probabilities and the probability in the other two columns are conditional probabilities.

Table 5-1

Probabilities of Oil Well Driller Problem

Low cost survey	oil present (P) $P(P) = .40$	$P(O_1/P) = .70$	$P(\bar{O}_1/P) = .30$
	no oil present ($\bar{P}$) $P(\bar{P}) = .60$	$P(O_1/\bar{P}) = .40$	$P(\bar{O}_1/\bar{P}) = .60$
High Cost survey	oil present (P) $P(P) = .40$	$P(O_2/P) = .80$	$P(\bar{O}_2/P) = .20$
	no oil present ($\bar{P}$) $P(\bar{P}) = .60$	$P(O_2/\bar{P}) = .30$	$P(\bar{O}_2/\bar{P}) = .70$

Possible Alternatives

The alternatives open to the oil well driller are
several. He can drill without a survey, with the low cost
survey, with the high cost survey or he may decide not to
drill at all. Which one of the four alternatives should he
choose? The decision tree in Figure 5-1 presents an analysis
of the problem. Let us assume that if he finds oil his net
payoff is $1,000,000 and if he finds no oil his cost is
$200,000. The cost of the two geophysical surveys are

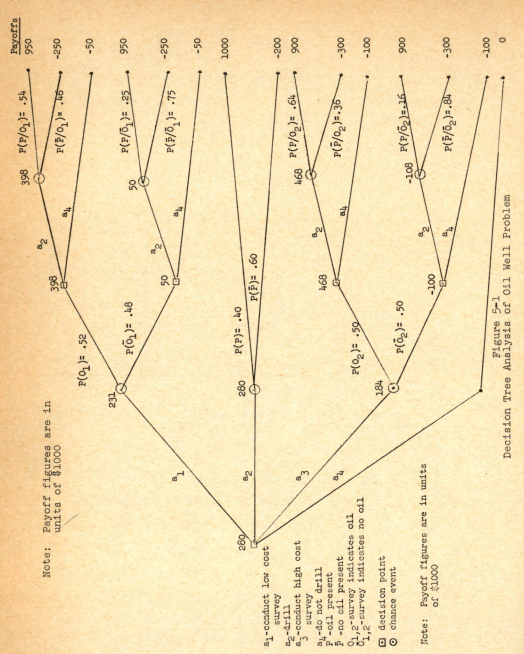

Payoffs

950
-250
-50
950
-250
-50
1000
-200
900
-300
-100
900
-300
-100
0

Note: Payoff figures are in
units of $1000

$P(P/O_1) = .54$
$P(\bar{P}/O_1) = .46$
$P(P/\bar{O}_1) = .25$
$P(\bar{P}/\bar{O}_1) = .75$
$P(\bar{P}) = .60$
$P(P/O_2) = .64$
$P(\bar{P}/O_2) = .36$
$P(P/\bar{O}_2) = .16$
$P(\bar{P}/\bar{O}_2) = .84$

398
50
468
-108

a_2
a_4
a_2
a_4
a_2
a_4
a_2
a_4

398
50
468
-100

$P(O_1) = .52$
$P(\bar{O}_1) = .48$
$P(P) = .40$
$P(O_2) = .50$
$P(\bar{O}_2) = .50$

231
280
184

a_1
a_2
a_3
a_4

280

a_1–conduct low cost
    survey
a_2–drill
a_3–conduct high cost
    survey
a_4–do not drill
P –oil present
$\bar{P}$ –no oil present
$O_{1,2}$–survey indicates oil
$\bar{O}_{1,2}$–survey indicates no oil

☐ decision point
⊙ chance event

Note: Payoff figures are in units
    of $1000

Figure 5-1
Decision Tree Analysis of Oil Well Problem

respectively $50,000 and $100,000.

The decision tree can be divided into four parts or stages. At the first stage, to the left, the four possible initial actions of drilling, not drilling and conducting either the low or high cost survey are presented. The "do not drill" action ends at the end of the first stage with a payoff of zero as indicated. The second stage is a chance stage. That is each one of the three action end points has either one of the two outcomes, there is oil or there is no oil present. Although the presence or absence of oil is only discovered in the drill action the same two conditions still hold true for the other cases.

At the end of stage two oil is struck or a dry hole is found. Hence stage two is the final one for the drill action. The appropriate payoffs are shown. The survey actions now proceed to the third stage and two alternatives are presented. One alternative is drill and the other alternative is do not drill. The do not drill action ends at the end of the third stage showing the respective costs of the low and high cost surveys as negative payoffs. The final stage shows the chance alternatives of oil and no oil with the respective payoffs.

Backward Induction

The next step in the decision tree analysis is called backward induction. Starting from the right we move backward

to the left in order to determine what is the best strategy
for the oil well driller. Looking at the two branches at
the upper right hand corner note that there is a payoff of
$950 with a probability of 0.54 and a negative payoff (or
cost) of $250 with a probability of 0.46. How the probabili-
ties are derived will be discussed below. The expected
value of these two payoffs is thus $398 as noted directly
above the upper chance node at the beginning of stage four.
Doing the same for the other three nodes at stage four expected payoffs are obtained of $50, $468 and -$108.

Working backward through the third stage note that
the oil driller had to choose between action a_2 and action
a_4. Based on the information we have at this point it is
expected that the oil well driller in all four cases would
select action a_2 because it provides a better payoff than
action a_4 in each instance. Hence at the beginning of stage
three the respective payoffs are $398, $50, $468 and -$100.

Stage two is another chance stage with different
probabilities. Finding the expected payoffs for each of the
three sets of branches we find respective payoffs of $231,
$280 and $184. At the first stage, being a decision stage,
the highest expected payoff of $280 would, of course, be
selected. This completes the backward induction phase and
the optimal (highest expected payoff) strategy for the oil
well driller is to have no surveys conducted and drill.

Bayesian Analysis[2]

The conditional probabilities, also called posterior probabilities, at the last stage, $P(P/O)$, $P(\bar{P}/O)$, $P(P/\bar{O})$, and $P(\bar{P}/\bar{O})$ are derived from Bayes' formula, using the prior probabilities $P(P)$ and $P(\bar{P})$ and the conditional probabilities listed in Table 5-1 as follows:

$$P(P/O_1) = \frac{P(O_1/P)\ P(P)}{P(O_1/P)\ P(P) + P(O_1/\bar{P})\ P(\bar{P})} = \frac{(.70)(.40)}{(.70)(.40)+(.40)(.60)}$$

$$= \frac{.28}{.52} = .54$$

and

$$P(P/\bar{O}_1) = \frac{P(\bar{O}_1/P)\ P(P)}{P(\bar{O}_1/P)P(P)+P(\bar{O}_1/\bar{P})P(\bar{P})} = \frac{(.30)(.40)}{(.3)(.4)+(.6)(.6)} = \frac{.12}{.48} = .25$$

and

$$P(P/O_2) = \frac{P(O_2/P)\ P(P)}{P(O_2/P)P(P)+P(O_2/\bar{P})P(\bar{P})} = \frac{(.80)(.40)}{(.8)(.4)+(.3)(.6)} = \frac{.32}{.50} = .64$$

and

$$P(P/\bar{O}_2) = \frac{P(\bar{O}_2/P)\ P(P)}{P(\bar{O}_2/P)P(P)+P(\bar{O}_2/\bar{P})P(\bar{P})} = \frac{(.20)(.40)}{(.2)(.4)+(.7)(.6)} = \frac{.08}{.50} = .16$$

---

[2]For a more detailed discussion of Bayesian analysis the reader is referred to J. Hirshleifer; "The Bayesian Approach to Statistical Decision-An Exposition," The Journal of Business, Vol. 24 (October 1961), pp. 471-89; R. Schlaifer: Probability and Statistics for Business Decisions, New York: Mc-Graw-Hill Book Company, 1959; and H. Raiffa and R. Schlaifer: Applied Statistical Decision Theory, Boston, Massachusetts: Harvard University, Division of Research, Graduate School of Business Administration, 1961.

The probabilities $P(\bar{P}/O_1)$, $P(\bar{P}/\bar{O}_1)$, $P(\bar{P}/O_2)$ and $P(\bar{P}/\bar{O}_2)$ are just equal to one minus the above probabilities or .46, .75, .36 and .84 respectively.

The probabilities in the third stage are found after the last stage probabilities have been determined with the exception, of course, of $P(P)$ and $P(\bar{P})$ which are known. To find the other second stage probabilities the following formula is used for $P(O_1)$, $P(\bar{O}_1)$, $P(O_2)$ and $P(\bar{O}_2)$.

$$P(O_1) = P(O_1/P)\ P(P) + P(O_1/\bar{P})\ P(\bar{P}) = .28 + .24 = .52$$

and

$$P(\bar{O}_1) = P(\bar{O}_1/P)\ P(P) + P(\bar{O}_1/\bar{P})\ P(\bar{P}) = .12 + .36 = .48$$

and

$$P(O_2) = P(O_2/P)\ P(P) + P(O_2/\bar{P})\ P(\bar{P}) = .32 + .18 = .50$$

and

$$P(\bar{O}_2) = P(\bar{O}_2/P)\ P(P) + P(\bar{O}_2/\bar{P})\ P(\bar{P}) = .08 + .42 = .50$$

The solution to the problem is also shown in tabular format in Table 5-2. Note that the maximum of the four payoffs is represented by action a_2. However, the information in Table 5-3 was obtained from the decision tree in Figure 5-1 by backward induction. Hence, the tabular method is not a replacement for the decision tree approach but only a means of checking the payoffs found directly by backward induction.

Table 5-2

Payoff Calculations for Oil Well Driller

Stages	Payoffs*	Probability	Expected Payoffs*
a_1-0-a_2-P/0	950	(.52)(.54)	267
a_1-0-a_2-$\bar{P}$/0	-250	(.52)(.46)	-60
a_1-$\bar{0}$-a_2-P/$\bar{0}$	950	(.48)(.25)	114
a_1-$\bar{0}$-a_2-$\bar{P}$/$\bar{0}$	-250	(.48)(.75)	$\underline{-90}$
		Expected payoff for action a_1	231
a_2-P	1000	(.40)	400
a_2-$\bar{P}$	-200	(.60)	$\underline{-120}$
		Expected payoff for action a_2	280
a_3-0-a_2-P/0	900	(.50)(.64)	288
a_3-0-a_2-$\bar{P}$/0	-300	(.50)(.36)	-54
a_3-$\bar{0}$-a_4	-100	(.50)	$\underline{-50}$
		Expected payoff for action a_3	184
a_4	0	(1.00)	$\underline{0}$
		Expected payoff for action a_4	0

*Payoffs and expected payoffs are in units of $1000.

A PLANT INVESTMENT DECISION

The rather simple decision tree problem discussed in the previous example of the oil well driller quickly grew

rather complex and the question arises how this method can
be applied in more complex investment situations. The an-
swer is that too much complexity should be avoided unless
additional complexity adds significantly to the improvement
of the analysis. However, the number of stages one would go
through is usually limited to four or six stages. For instance, a marketing research study for an investment problem
may consist of a market survey and a test market sequentially
thus producing a six-stage decision tree analysis. Analyses
exceeding six stages are rather rare and if six stages are
exceeded the decision tree becomes rather complex.

Investment Alternatives

To illustrate a different six-stage investment decision solution an investment analysis will be presented where
the decision makers must decide between several alternatives.
The alternatives are to build a large plant, to build a small
plant which subsequently may be enlarged to a large plant
but at a higher cost than if the large plant were built at
once, and to have a research study made on the basis of
which a large or a small plant would be built.

The analysts making the study have acquired a fairly
large amount of information on which to base their final
decision. From the marketing department they have been provided with three sales estimates and their respective probabilities. A probability of 0.50 has been assigned to a

demand which would require the large plant; for a demand which could be satisfied by a small plant a probability of 0.30 has been assigned. There also is a probability of 0.20 that demand will initially be high, thus necessitating a large plant but demand would subsequently decline so that a small plant would suffice.

Research Study

The research study if necessary is to cost $300,000 and will be performed by a team which can be expected to supply forecasts, based on historical performance, which have the following probabilistic properties. If a large plant is required to satisfy demand, there is a 70% chance that the research team will report a high demand. If demand is such that it can be supplied by a small plant, then there is a 10% chance that the research team will report a high demand. If there is going to be a declining demand starting with an initial high demand there is a 50% chance that the research team will forecast a high demand.

Table 5-3 presents a summary of the above probabilities. Note that the three probabilities in the first column are simple probabilities and the others are conditional probabilities.

Estimated Costs and Revenues

The cost of building a large plant or a small plant
is five million and three million dollars respectively. To
expand a small plant to a large plant costs four million
dollars. The net discounted cash profits for the various
alternatives, excluding the costs of the plants or the

Table 5-3

Probability Values of Plant Investment Problem

Large market (L)		
P(L) = .50	P(H/L) = .70	P($\bar{H}$/L) = .30
Declining market (D)		
P(D) = .20	P(H/D) = .50	P($\bar{H}$/D) = .50
Small market (S)		
P(S) = .30	P(H/S) = .10	P($\bar{H}$/S) = .90

Note: H - research team forecasts high demand
       $\bar{H}$ - research team forecasts low demand

research study, are ten million dollars for low demand, fif-
teen million dollars for declining demand, if a large plant
is available to satisfy that demand, and twenty million
dollars for high demand again of course assuming that the
large plant is built. An additional net discounted cashflow of one million dollars is earned if demand is high but
only a small plant is available. The additional cashflow

is earned from working the small plant more intensively
through the use of overtime and round-the-clock operation.
The net discounted cashflows at the right of the decision
tree in Figure 5-2 are found by taking the above figures and
subtracting the cost of the plants and the research study
if appropriate. Table 5-4 lists the net payoffs for each branch.

The Analysis

If the reader understood the oil well driller decision tree analysis he should have little trouble following
the analysis for the investment problem. The twenty-four
end points at the last stage are all chance events whose
respective probabilities are either known or have been calculated by Bayes' or other formulas. Moving downward note
that the first three probabilities are known and listed in
Table 5-3. The second three probabilities are posterior
probabilities found by Bayes' formula and with probability
values given in Table 5-3. The formula for the first posterior probability value is,

$$P(L/H) = \frac{P(H/L)\ P(L)}{P(H/L)P(L)+P(H/D)P(D)+P(H/S)P(S)} =$$

$$\frac{(.70)(.50)}{(.7)(.5)+(.5)(.2)+(.1)(.3)} = \frac{.35}{.48} = .73$$

The other two are similarly calculated.

Table 5-4 Branch Payoffs for Investment Problem

	a_1	a_3	$a_3 - a_4$	$a_3 - a_5$
L	20-15=15	11-3=8	20-3-4=13	11-3=8
D	15- 5=10	10-3=7	15-3-4= 8	10-3=7
S	10- 5= 5	10-3=7	10-3 = 7	10-3=7

Notes: Subtract 0.3 from the above figures if the research study is performed.
Entries are in millions of dollars.

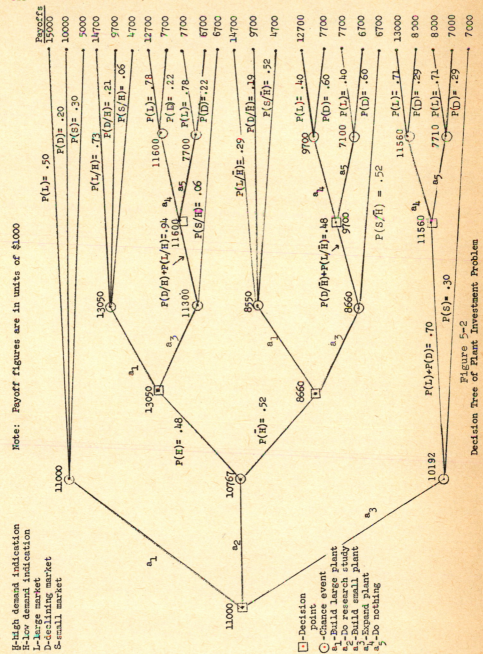

H-high demand indication
H-low demand indication
L-large market
D-declining market
S-small market

Note: Payoff figures are in units of $1000

☐ -Decision
    point
⊙ -Chance event
a₁-Build large plant
a₂-Do research study
a₃-Build small plant
a₄-Expand plant
a₅-Do nothing

Figure 5-2

Decision Tree of Plant Investment Problem

The probability values of those branches that only cover the last stage, P(L) and P(S) are found by the ratios of the probabilities in the preceding fourth stage. Thus the formula for the seventh branch from the top is,

$$P(L) = \frac{P(L/H)}{P(D/H) + P(L/H)} = \frac{.73}{.73 + .21} = \frac{.73}{.94} = .78$$

The probability for P(D) is of course .22 and all other P(L)'s and P(D)'s are similarly determined.

The two probabilities at the second stage P(H) and $P(\bar{H})$ are determined by the following formulas,

$$P(H) = P(H/L)P(L) + P(H/D)P(D) + P(H/S)P(S)$$

$$= (.7)(.5) + (.5)(.2) + (.1)(.3) = .48$$

and

$$P(\bar{H}) = P(\bar{H}/L)P(L) + P(\bar{H}/D)P(D) + P(\bar{H}/S)P(S)$$

$$= (.3)(.5) + (.5)(.2) + (.9)(.3) = .52$$

The optimal investment decision is to build the large plant. The solution is found by backward induction. For every set of chance events the expected payoff is found starting from the right side of the decision tree. The expected payoff thus calculated is shown directly above each chance node. For each set of actions the highest expected payoff is selected and "moved back." The highest payoff is shown directly above the decision point. Working

all the way back to the left of the decision tree you may
observe that the highest expected net discounted cash flow
is 11000 or rather eleven million dollars since all dollar
values in Figure 5-2 are shown in units of one thousand dollars.

In addition to helping management in deciding among
alternative actions the decision tree also reveals some
interesting information which may help management in deciding
to follow an alternative course of action. For instance
note that the next best decision to have a research study
made has a net expected cash profit payoff which is only
slightly over two percent lower than the optimal decision.
Even the worst of the three initial actions still promises
a net expected cash profit which is only about eight percent
less than the optimum decision. The above information could
be especially valuable if the firm for which the decision
is made has problems in raising the five million dollars
required for the large plant. Even the small plant at a
cost of only three million dollars initially still provides
approximately 92 percent of the net expected cash profit
of the large plant. Hence, one can see that decision trees
do more than just select the optimum or maximum profit
solution; they also provide information on which alternative,
not necessarily optimal, decisions can be made.

A SEQUENTIAL INVESTMENT DECISION

The title sequential decision problem is chosen
for this section because two sequential tests will be
evaluated instead of just one for the oil well driller
problem and a mixed test for the plant investment problem.
The plant investment problem, of course, had two sequential
alternatives in building a small plant which could subsequently be followed by a plant expansion. However, in both
previous problems only one research study or geophysical
test was performed before the decision on whether to proceed
or not was made. The oil well driller problem had the op-
tion of two alternative tests but both were in parallel and
not in series (i.e. sequential).

A Consumer Panel and a Test Market

The investment problem to be discussed here involves
a decision on whether to launch a new consumer product with
all its concomitant investment requirements of production
operation, marketing, sales force, etc. The first optional
test is a consumer test panel which has a relatively low
cost of $100,000. The second test which follows the con-
sumer test panel is a consumer test market with a relatively
high cost of $400,000. If the product is successful the
net expected discounted cash profit is one million dollars
and if unsuccessful the net expected discounted cash cost

is \$200,000. From the net expected discounted cash profits
must of course be subtracted any costs which are incurred
as part of the test panel or test market.

Management has investigated what the probabilities
of success and failure are without any marketing research
and has arrived at a probability of success of 0.60 and a
probability of failure of 0.40. The conditional probabilities of the test panel are presented in Table 5-5 and of the
test market in Table 5-6. To calculate the posterior probabilities for the test market with Bayes' formula the posterior probabilities of the test panel are used as the
prior probabilities for the test market. This is a rational
procedure because the original probabilities of success and
failure have been modified by the results of the test panel
and any information gained from the test panel should be
incorporated in subsequent analysis.

Table 5-5

Probability Values for Consumer Test Panel

Success (S) $P(S) = .60$	$P(H/S) = .70$	$P(\bar{H}/S) = .30$
Failure (F) $P(F) = .40$	$P(H/F) = .25$	$P(\bar{H}/F) = .75$

Note: H - test panel predicts high demand
$\bar{H}$ - test panel predicts low demand

Table 5-6

Probability Values for Consumer Test Market

Success (S)		
$P(S/H) = P(S_1) = .81$	$P(LH/S_1) = .80$	$P(\bar{L}H/S_1) = .20$
Failure (F)		
$P(F/H) = P(F_1) = .19$	$P(LH/F_1) = .15$	$P(\bar{L}H/F_1) = .85$
Success (S)		
$P(S/\bar{H}) = P(S_2) = .38$	$P(L\bar{H}/S_2) = .65$	$P(\bar{L}\bar{H}/S_2) = .35$
Failure (F)		
$P(F/\bar{H}) = P(F_2) = .62$	$P(L\bar{H}/F_2) = .10$	$P(\bar{L}\bar{H}/F_2) = .90$

Note: LH - test market predict large market following
            positive test panel results

   $\bar{L}H$ - test market predict small market following
            positive test panel results

   $L\bar{H}$ - test market predict large market following
            negative test panel results

   $\bar{L}\bar{H}$ - test market predict small market following
            negative test panel results

Tree Analysis

        The decision tree is shown in Figure 5-3. It is a
six-stage model. The first stage presents a three-alterna-
tive decision: launch the product, do not launch the product or commission a consumer test panel. The second stage
is a chance stage. At the third stage a decision can be
made to launch the product, do not launch the product or
commission a test market. At the fourth stage the

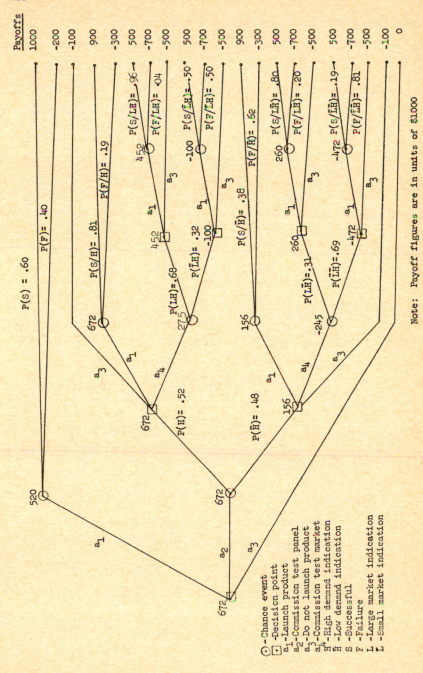

Payoffs (in units of $1000)

	1000
	-200
	-100
	900
	-300
	500
	-700
	-500
	500
	-700
	-500
	900
	-300
	500
	-700
	-500
	500
	-700
	-500
	-100
	0

P(S) = .60
P(F) = .40
P(S/H) = .81
P(F/H) = .19
P(S/LH) = .96
P(F/LH) = .04
P(S/L̄H) = .50
P(F/L̄H) = .50
P(LH) = .68
P(L̄H) = .32
P(S/H̄) = .38
P(F/H̄) = .62
P(S/LH̄) = .80
P(F/LH̄) = .20
P(S/L̄H̄) = .19
P(F/L̄H̄) = .81
P(LH̄) = .31
P(L̄H̄) = .69
P(H) = .52
P(H̄) = .48

520
672
672
672
672
275
452
452
156
156
260
260
-100
-245
-472
-472

Note: Payoff figures are in units of $1000

⊙ - Chance event
▢ - Decision point
a_1 - Launch product
a_2 - Commission test panel
a_3 - Do not launch product
a_4 - Commission test market
H - High demand indication
H̄ - Low demand indication
S - Successful
F - Failure
L - Large market indication
L̄ - Small market indication

Figure 5-3

Decision Tree for Consumer Product Investment Problem

posterior probabilities are calculated for $P(S/H)$, $P(F/H)$, $P(S/\bar{H})$ and $P(F/\bar{H})$. The fifth stage is another decision stage where the two alternatives are to launch or not to launch the product. The sixth and final stage is a chance stage and the posterior probabilities $P(S_1/LH)$, $P(F_1/LH)$, etc. are calculated.

To calculate the set of posterior probabilities following the test panel results the following formulas are used,

$$P(S/H) = \frac{P(H/S)\ P(S)}{P(H/S)P(S)+P(H/F)P(F)} = \frac{(.7)(.6)}{(.7)(.6)+(.25)(.4)} = \frac{.42}{.52} = .81$$

and

$$P(S/\bar{H}) = \frac{P(\bar{H}/S)\ P(S)}{P(\bar{H}/S)P(S)+P(\bar{H}/F)P(F)} = \frac{(.3)(.6)}{(.3)(.6)+(.75)(.4)} = \frac{.18}{.48} = .38$$

The probability, $P(F/H)$ is just $1-P(S/H)$ or .19 and $P(F/\bar{H})$ is $1-P(S/\bar{H})$ or .62.

From the above probabilities $P(H)$ and $P(\bar{H})$ in the second stage of the decision tree can be determined as follows,

$$P(H) = P(H/S)\ P(S)+P(H/F)\ P(F) = .52$$

Probability $P(\bar{H})$ is just $1-P(H)$ or .48.

To calculate the posterior probabilities in the last stage of the decision tree probability information from Table 5-5 will be used in the following Bayes' formulas,

$$P(S_1/LH) = \frac{P(LH/S_1) \; P(S_1)}{P(LH/S_1)P(S_1)+P(LH/F_1) \; P(F_1)} = \frac{(.80)(.81)}{(.80)(.81)+(.15)(.19)}$$

$$= \frac{.648}{.677} = .96$$

and

$$P(S_1/\bar{L}H) = \frac{P(\bar{L}H/S_1) \; P(S_1)}{P(\bar{L}H/S_1) \; P(S_1)+P(\bar{L}H/F_1)P(F_1)} = \frac{(.20)(.81)}{(.20)(.81)+(.85)(.19)}$$

$$= \frac{.162}{.324} = .50$$

and

$$P(S_2/L\bar{H}) = \frac{P(L\bar{H}/S_2) \; P(S_2)}{P(L\bar{H}/S_2) \; P(S_2)+P(L\bar{H}/F_2) \; P(F_2)} = \frac{(.65)(.38)}{(.65)(.38)+(.10)(.62)}$$

$$= \frac{.247}{.309} = .80$$

and

$$P(S_2/\bar{L}\bar{H}) = \frac{P(\bar{L}\bar{H}/S_2) \; P(S_2)}{P(\bar{L}\bar{H}/S_2) \; P(S_2)+P(\bar{L}\bar{H}/F_2) \; P(F_2)} = \frac{(.35)(.38)}{(.35)(.38)+(.90)(.62)}$$

$$= \frac{.133}{.691} = .19$$

and

$$P(F_1/LH) = 1 - P(S_1/LH) = .04$$

and

$$P(F_1/\bar{L}H) = 1 - P(S_1/\bar{L}H) = .50$$

and

$$P(F_2/L\bar{H}) = 1 - P(S_2/L\bar{H}) = .20$$

and

$$P(F_2/\bar{L}\bar{H}) = 1 - P(S_2/\bar{L}\bar{H}) = .81$$

To find the probabilities in the fourth stage of the decision tree the following formulas are applied,

$$P(LH) = P(LH/S_1) \ P(S_1) + P(LH/F_1) \ P(F_1) = .68$$

and

$$P(\bar{L}H) = P(\bar{L}H/S_2) \ P(S_2) + P(\bar{L}H/F_2) \ P(F_2) = .32$$

and

$$P(L\bar{H}) = P(L\bar{H}/S_2) \ P(S_2) + P(L\bar{H}/F_2) \ P(F_2) = .31$$

and

$$P(\bar{L}\bar{H}) = P(\bar{L}\bar{H}/S_2) \ P(S_2) + P(\bar{L}\bar{H}/F_2) \ P(F_2) = .69$$

The method of backward induction is applied to the decision tree and the net expected discounted cash profits are calculated for each node of the decision tree as indicated in Figure 5-3. Based on the calculations one may observe that the optimum (maximum expected discounted cash profit) strategy is to commission a consumer test panel at a cost of $100,000 and then launch the product.

However, if it is decided to proceed as described above management can, of course, re-evaluate its decision following the test panel completion. Additional information is available and a complete re-analysis may be required.

ADDITIONAL COMMENTS

In the above examples knowledge of discrete probabilities has been assumed. However, frequently probabilistic knowledge is only available in other forms. For instance,

a marketing organization may forecast sales of at least a
million dollars with certainty, sales of at least one and a
half million dollars with a probability of .50 and sales of
two million dollars with a probability of .10. How can this
information be translated into discrete probabilistic form?
The probabilities provided are cumulative probabilities
and to simplify them to discrete simple probabilities the
smallest cumulative probability must be subtracted from the
next cumulative probability, etc. The result is that a
reasonable probability assignment would be the following:

Sales	Probability
1,000,000	.50
1,500,000	.40
2,000,000	.10

Another type of forecast may be of the following
form: Sales of one to two million dollars are twice as
probable as sales of two to three million dollars and the
latter are again twice as probable as sales of three to
four million dollars. How can this information be used to
assign probabilities? Well, there are three center points
which will each have to be assigned simple discrete probabilities. These three center points are one and a half
million, two and a half million and three and a half million
dollars of sales, with probabilities of 4/7, 2/7 and 1/7
respectively.

EXERCISES

1. Develop a decision tree analysis for launching a new
   product with three alternative advertising campaigns.
   Campaign A cost one million dollars, campaign B cost
   two million dollars and campaign C costs one million
   dollars now and the present value equivalent of two
   million dollars half a year from now. The probabilities
   of net discounted profits over the life of the product
   are as shown below.

Net Payoff in	Probability for Campaign		
Million Dollars	A	B	C
15	.05	.10	.15
12	.10	.15	.20
10	.15	.20	.25
8	.20	.25	.30
6	.30	.20	.10
4	.20	.10	0

2. A medical doctor must decide to perform a surgical procedure on a patient, to apply a therapeutic procedure
   or to have the patient undergo a medical test. The
   test will indicate whether surgery or therapy are the
   required medical approaches to cure the patient. If the
   wrong medical approach is used, i.e. if therapy is
   applied while surgery is needed or vice versa then
   the patient is put in danger. Similarly delaying the
   medical approaches in order to apply the medical test

also may endanger the patient. The subjective probability that surgery (S) or therapy (T) are needed are
0.40 and 0.60 respectively. If therapy is needed the
test will have a low (L) reading with a probability of
0.75 and a high reading with a probability of 0.25.
On the other hand if surgery is required the test will
have a low reading with a probability of 0.10 and a
high reading with a probability of 0.90. The utility
payoffs are 100 units if the proper medical procedure
is performed and 20 units if the incorrect medical procedure is performed. If the test is performed 10
utility units are subtracted and the payoffs become
90 and 10 utility units respectively. Analyze the above
problem with a decision tree.

3. A company wants to buy 50,000 pieces of a given part.
    A newly-developed product is currently available which
    is presumably more reliable than the old type, which
    had a 98% reliability history. The new part costs
    $250 more for 50,000 pieces. If an item is defective,
    it costs the company $1.00 for rework. Based on the
    experience of other companies, which use the new part,
    the following probabilities are assigned:

Reliability of new part	Probability
0.99	0.6
0.98	0.3
0.97	0.1

(a) Should the company switch over to the new part?

(b) A sample of 20 pieces of the new part produces one defective. What should the company do now?

4. Servospace, a space-age technology supplier of servo-systems adjusts one of its most complicated products called the Hilo under simulated field conditions at its plant prior to shipment. Following the adjustment, each Hilo is shipped to the purchaser for incorporation in the purchaser's system. However, during shipment to the purchaser the Hilo occasionally gets out of alignment. The purchaser upon receipt of each Hilo, checks it out with a simpler piece of testing equipment which is not as accurate as the alignment process performed at the plant under simulated field conditions. Fortunately, historical data has indicated that the Hilo purchaser's test equipment will discover that the Hilo is out of alignment, if it actually is out of alignment, with a probability of 0.90. If the Hilo is not out of alignment, the purchaser's test equipment will so indicate with a probability of 0.95. Fortunately, Servospace has kept track of all Hilos shipped out and has found that there is a ten percent chance of a Hilo going out of alignment during shipment.

    The cost of correcting a misaligned Hilo in the purchaser's system costs Servospace $40,000. This occasional high cost item can be avoided by performing an alignment test at the purchaser's plant under simulated field conditions. The cost of this to Servospace, however, amounts to $1,000, whereas the simple test now performed by the purchaser only costs $100. Servospace has just encountered a $40,000 charge for correcting a misaligned Hilo and is meeting to decide whether it should continue its practice or change to a $1,000 field test on each Hilo. What is your recommendation?

5. For many decision problems encountered in real world situations, the conditional and other probabilities cannot always be derived from historical data. In some cases not even directly related historical data is available and the decision maker must use subjective probabilities. In these two cases, can the decision tree framework still be used or does it become worthless? Discuss.

6. Catalytic Chemical, Inc. is faced with the prospect of having to choose one of three possible processes for producing a new chemical industrial product. The selected process will definitely be used for one year after which time the entire product, production and marketing, will be re-examined. Hence, any investments in fixed assets will be considered as an expense for the year. The production level, cost and investment requirements are shown below. For processes B and C current process facilities can be used and incremental investment is therefore nil. The finished product can be sold for $1 per pound. Assume that the firm decides to select one process. Additional experimentation of ten trials

Processes	A	B	C
Production in batches	110	100	90
Cost per batch, material and labor	$500	$600	$700
Investment	$7500	0	0
Pilot plant yield in lbs. per batch from ten trials for each process	950	1000	1150
Cost for further experimentation per trial	$100	$110	$120

for each process is possible if deemed necessary. Assume that the outcome of the experimentation is expected to provide the results shown below. These expected results are based on a planned modification in all three processes instituted following the previous experimentation. Hence, this modification of the process has doubled the alternatives available before.

Processes	A		B		C	
	Yield	Probability	Yield	Probability	Yield	Probability
Pilot plant yield in lbs. per batch from ten trials for each process	900	.30	900	.50	900	.20
					1200	.50
	1000	.70	1100	.50	1300	.30

Draw up the decision tree and determine which process Catalytic Chemical should select.

7. Management Consultants Inc. is presented with a decision problem by one of its clients. The client is trying to decide whether or not to launch a new product. The new product appears quite attractive from a profit point of view. However, because of the uncertainties of competitor reactions the problem is so complex that a detailed analysis is required. The probability of a competitor entering the market with a similar product is 0.80. The client is thinking of three pricing strategies, high, medium and low. In response, the client's competitor or competitors will also price their product high, medium or low as shown in the following probability table. The condition 1 profits in money units associated with each one of the competitor's reactions are also shown in the table. If no

Client Pricing	Competitor Pricing	Probability	Conditional Profit
	High	0.4	70
High	Medium	0.5	30
	Low	0.1	−20
	High	0.1	80
Medium	Medium	0.6	40
	Low	0.3	−30
	High	0.1	80
Low	Medium	0.2	45
	Low	0.7	−40

competitor enters the market the client expects conditional profits of 80, 60 and 10 for high, medium and low price strategies respectively. If the client decides to launch the new product management of the client feels that it does so at an opportunity cost of 20 money units. Analyze the problem for the client and suggest a course of action.

8. If the client of Management Consultants in exercise 7 is unsure of its probability estimates and if better probability estimates can be obtained

at a cost of 5 money units would you recommend the expenditure?

9. A European brewery is planning to import and distribute its beer into a large U.S. metropolitan area. It estimates that the probability of a successful introduction and acceptance in the area is 0.75. It has also estimated that the present values of marginal profit of the decision are as shown below.

Decision	Successful	Unsuccessful
Introduce	$800,000	$-10,000
Do Not Introduce	0	0

Before making the decision the brewery decides to test the market by a sample survey. This sample survey will be made by a local marketing agency and will cost $3000. The agency has made many similar surveys and it presents the following conditional probabilities for the survey.

P (Positive response/successful introduction) = .80

P (Negative response/unsuccessful introduction) = .30

Use the decision tree concept to analyze this problem. Should the brewery introduce its beer if the response to the survey is positive?

10. Paoloni Plastics is considering expansion in the plexiglass industry and has an opportunity to acquire a firm in the plexiglass industry. However, management is still concerned with the future of plexiglass. Plexiglass industry publications have predicted the following future of plexiglass.

P (Very rapid growth) = .3

P (Moderate growth) = .5

P (No growth)   = .2

The attractiveness of the acquisition is to a large extent dependent on the growth prospects of plexiglass. Because of the high price Paoloni has to pay for the plexiglass firm it would not be a desirable acquisition if there was no growth. The following table presents the net present value of acquiring the plexiglass firm.

Decision	Very Rapid Growth	Moderate Growth	No Growth
Acquire	$5,000,000	$1,000,000	−2,000,000
Do Not Acquire	0	0	0

Paoloni decides to engage a consulting firm for a fee of $3000 to study the future of the plexiglass firm and industry. The consulting firm has made many of such similar studies and has the following record.

P (Recommend acquisition/very rapid growth) = .7

P (Recommend acquisition/moderate growth) = .2

P (Recommend acquisition/no growth)   = .1

Analyze the problem with a decision tree. Should Paoloni acquire the plexiglass firm if the consultants so recommend?

CHAPTER 6.--CAPITAL INVESTMENT ANALYSIS BY SIMULATION[1]

Whenever a capital investment decision has to be made the problem arises on which criterion the decision has to be based. Once the criterion has been chosen, by some arbitrary method, the chosen criterion is then commonly used to make future capital investment decisions. However, the calculations required for comparing capital investments on the basis of one criterion usually only require a relatively small extension in order to compare the capital investments on the basis of several criteria.

DECISION CRITERIA

This chapter will review six objective and one subjective capital investment decision criteria, consisting of two rate of return criteria, two payback criteria, two discounted cash flow criteria, and a subjective utility criterion. It is assumed that a subjective utility can be obtained.

To illustrate how the seven decision criteria compare, two hypothetical capital investment proposals are presented. It is assumed that one of the investment proposals has to be selected on the basis of the decision criteria. It will

---

[1]This chapter is based on C. Carl Pegels: "A Comparison of Decision Criteria for Capital Investment Decisions," The Engineering Economist, Vol. 13 (September 1968), pp. 211-20.

be shown that, what appears to be the best proposal, is in essence inferior to the other proposal. The two hypothetical capital investment proposals will be presented in the form of probability distributions of input variables, such as sales rate, sales growth, product price, capital investment required, and so forth. A simulation experiment will then be used to sample from these input variable distributions and using the sampled values, the estimators of the output variable distribution parameters will be calculated. The output data will also be tested to determine if the distributions of the output variables conform to common analytic distributions.

Review of Objective Decision Criteria

The six objective decision criteria will be reviewed in this section. In the next section, the subjective decision criterion will be discussed.

The first decision criterion is the internal rate of return which is determined by setting present value (PV) equal to 0 and solving for internal rate of return (r) in the equation:

$$PV = 0 = \sum_{i=1}^{n} \frac{x_i}{(1+r)^{i+1}} - I$$

where PV = present value or net discounted cash flow for
life of investment, assuming zero salvage value.

> I = total cash outlay for capital investment, assumed
> to have been made as a lump sum at the beginning
> of year 0.
>
> n = number of years facility or equipment is expected
> to produce cash flows after year 0.
>
> x_i = cash flow in period i, assumed to be received as
> a lump sum at the end of period i.

Baldwin[2] proposed an alternative rate of return method, here identified as the Baldwin rate of return method, because with the regular internal rate of return method the future receipts and payments are reduced to their present value by discounting them at the same rate as that which the proposed investment is estimated to provide. In other words, management assumes that, for the period between the base point and the time when the funds are spent or collected, the funds are, or could be, invested at the rate of return being calculated for the proposal.

The Baldwin rate of return is obtained by solving for (R) in the equation:

$$R = \left[\frac{\sum_{i=1}^{n} x_i (1+r_o)^{n-i}}{I} \right]^{\frac{1}{n+1}} - 1,$$

where r_o = the interest or earnings rate that the organization

---

[2] R.H. Baldwin, "How to Assess Investment Proposals," Harvard Business Review, May-June 1959, pp. 98-99.

on the average earns on its assets. A more accurate rate of return is provided by the Baldwin method than by the straight internal rate of return method. If r_o and R are alike, a coincidence, only then would the two methods provide the same result.

The following two decision criteria are related to the time required to earn back the capital investment in the proposals. The first is the non-discounted pay-out time in years (n). It can be obtained by solving for n in the equation:

$$I = \sum_{i=1}^{n-1} x_i$$

The other pay-out criterion is the discounted pay-out time in years (n), and can be obtained by solving for n in the equation:

$$I(1+r_o)^{n+1} = \sum_{i=1}^{n} x_i \ (1+r_o)^{n-i}$$

The last two objective decision criteria are related to discounted cash flows. The first of these is the discounted cash flow over the life of the investment. It will be identified as PV for present value. It can be found for given r_o as follows:

$$PV = \sum_{i=1}^{n} \frac{x_i}{(1+r_o)^{i+1}}$$

The other discounted cash flow criterion is discounted cash flow for an m-year period (PV_m). The m-year period would presumably be reasonably short, and, of course, would be shorter than the life of the investment. It is obtained for given r_o as follows:

$$PV_m = \sum_{i=1}^{m-1} \frac{x_i}{(1+r_o)^{i+1}}$$

Subjective Utility as a Decision Criterion

The subjective utility criterion to be used in this chapter will be subjective utility for money or cash. However, the problem that one still has to face is how to obtain utility functions for future years cash flow. Bierman, et.al.[3] discuss how to obtain a utility function for current period money. Furthermore, Grayson[4] and others have shown how a utility function for the firm and for individuals in the firm is constructed. However, nearly always the utility function is for the present or for a current period. But if a utility function for future years cash flow could be

---

[3] H. Bierman, Jr., L.E Fouraker and R.K. Jaedicke, Quantitative Analysis for Business Decisions. Homewood, Illinois: Richard D. Irwin, Inc., 1967, chapter 7.

[4] C.J. Grayson, Jr., Decisions Under Uncertainty. Boston, Massachusetts: Harvard University, Graduate School of Business Administration, 1960, chapter 10.

obtained, it could be applied to determine the total expected utility of an investment over its life.

Bierman, et.al., in their discussion of a common stock investment problem, propose the use of discounted utility, and the combining of the utility functions for different years. They point out, however, that there are theoretical problems in using that approach.

The author decided to stay with current practices and not use discounted utilities. However, if a firm could determine the utility of an investment and compare it with the utility of other investments, it would be of great help, because it would be a more meaningful addition to the criteria that have been applied in the past, such as rate of return, present value, years to pay-out, and so forth.

Cash flow, and especially cash flow in the years immediately following the commitment to an investment, are very important to the firm. If cash inflow is light during the first few years after an investment is made, the firm's cash balance may become too low and it may not be able to act as freely if a higher cash inflow from an investment were available. In other words, most firms are bound to have a higher utility for cash flow during the several years following the investment commitment than from the later years. On the basis of the above reasoning, it was decided to use as decision criterion the firm's utility for the discounted cash flow for the first m years after the investment commitment was made.

SIMULATION EXPERIMENT

The simulation experiment used is a sampling procedure whereby complicated expressions involving several probability distributions are evaluated. It consists of simulating an experiment to determine the probability distributions of populations of output variables by the use of random sampling applied to the input variables that will determine the output variables.

Hess and Quigley[5] applied the simulation method quite successfully to arrive at a distribution of rates of return for a $10,000,000 chemical plant expansion. Using a larger number of input variables Hertz[6] used the same method to arrive at a rate of return distribution for a proposed investment at a very low computer cost.

As was discussed before, in addition to rate of return, management should also investigate discounted cash

[5]S.W. Hess and H.A. Quigley, "Analysis of Risk in Investments Using Monte Carlo Techniques," Chemical Engineering Progress Symposium Series 42: Statistics and Numerical Methods in Chemical Engineering, New York: American Institute of Chemical Engineering, 1963, p.55.

[6]D.B. Hertz, "Risk Analysis in Capital Investment," Harvard Business Review, January-February 1964, pp. 95-106.

flow, years to pay-out and other decision criteria. With
the proper input variables the above-mentioned criteria,
and their respective probability distributions are available
at only slightly increased cost.

It is realized that the six decision criteria mentioned
above are not completely independent, but since they may be
obtained at only slightly increased cost, their availability
to management will further facilitate the capital investment
decision process.

How, then, can this simulation method be applied? To
illustrate this, two hypothetical investment decision problems were designed with the following significant input
variables: first year annual sales; annual sales growth
rate, as a percentage of first year sales; life of facility
in years; variable cost per unit in first year; selling
price per unit in first year; annual fixed cost; initial
investment required; change in selling price per unit from
year one to year two, to year three, and so forth; and
change in variable cost per unit from year one to year
two, to year three, and so forth.

It was assumed that management had estimated a
probability distribution[7] including its parameters for each
input variable as shown in Table 6-1.

_____

[7]For a reference on probability distributions please refer to an introductory probability and statistics textbook or to J.E. Freund, Mathematical
Statistics, Englewood Cliffs, N.J.: Prentice-Hall, Inc., 1962.

Table 6-1

Distributions and Parameters of Input Variables

Input Variable	Estimated Distribution	Parameter Values	
		Proposal A	Proposal B
First year annual sales in millions of units	Normal	μ = 1.10 σ = .10	μ = 1.00 σ = .10
Annual sales growth (a percentage of first year sales)	Normal	μ = 4.00 σ = .50	μ = 2.75 σ = .50
Life of facility in years	Normal	μ = 12.00 σ = .75	μ = 9.00 σ = .75
Variable cost per unit in first year in dollars	Normal	μ = 25.00 σ = 2.00	μ = 25.00 σ = 2.00
Selling price per unit in first year in dollars	Normal	μ = 42.00 σ = 2.00	μ = 43.00 σ = 2.00
Annual fixed cost in millions of dollars	Normal	μ = 9.00 σ = .70	μ = 9.00 σ = .70
Initial investment required in millions of dollars	Normal	μ = 19.00 σ = 2.00	μ = 16.00 σ = 2.00
Change in selling price and variable cost per unit in dollars (from year one to other years)	Uniform	α = -1 β = +3	α = -1 β = +3

Using the values, that the input variables assume for each simulation run, it is quite straight forward to determine the annual cash flow during the life of the investment, the various rates of return, the payback, and whatever other decision criterion that is desired.

Several assumptions have been made, one of which is the independence of sales volume and price. This assumption would not be realistic if the firm under study were the only or a major supplier of the product.

The simulation experiment was programmed to be executed on a computer. A total of 1000 simulation runs were made to obtain estimates of the probability distributions and the parameters of the six output variables, which represent the six objective decision criteria.

Table 6-2 shows the results of the simulation runs. The estimated underlying probability distributions of each output variable are listed for each case. The normality tests suggested by Hartley and Pearson[8] were used.

CASH FLOW UTILITY

The subjective utility criterion as discussed before is based on discounted cash flow. Discounted cash flow is used because it is assumed that the firm can put the cash flow to work at an average rate (r_o) for the firm. This average rate is assumed to be 10% in the example. The original investment is made at the beginning of period 0 and the annual cash flows are assumed to be received at the ends of the respective years. The period covered consists

---

[8] E.S. Pearson and H.O. Hartley, Biometrika Tables for Statisticians, Vol. 1, Cambridge: University Press, 1958, p. 61.

Table 6-2

Distributions and Parameters of Output Variables

Output Variables	Estimated Distribution	Estimated Values of μ and σ	
		Proposal A	Proposal B
Internal rate of return (percent)	Normal	$\mu = 42.2$ $\sigma = 11.4$	$\mu = 42.5$ $\sigma = 13.0$
Baldwin rate of return for $r_o = .10$ (percent)	*	$\mu = 23.4$ $\sigma = 3.8$	$\mu = 24.9$ $\sigma = 5.1$
Non-discounted payout time in years	#	$\mu = 3.22$ $\sigma = 1.23$	$\mu = 3.06$ $\sigma = 1.15$
Discounted payout time in years	$	$\mu = 4.15$ $\sigma = 4.97$	$\mu = 3.86$ $\sigma = 3.73$
Present value for $r_o = .10$ in millions of dollars	Normal	$\mu = 67.9$ $\sigma = 29.2$	$\mu = 42.7$ $\sigma = 20.9$
Net discounted cash flow for 4 years and for $r_o = .10$ in millions of dollars	Normal	$\mu = 12.3$ $\sigma = 11.2$	$\mu = 12.4$ $\sigma = 10.3$

* approaches the normal distribution
# approaches the lognormal distribution
$ neither normal nor lognormal

of four years. This period can be considered as a current short range planning horizon and the current utility of cash flow for the four-year period can be used. It is realized that the utility functions of the firm may change from year to year, but in order to make a decision on whether to invest now, or not at all, the current utility function is the only one available.

Since the example is a hypothetical one, the author has assumed the following utility function for discounted cash flow for the first four years:

$$U(x) = a[1 - \exp(-bx-c)]$$

The analytic form of the distributions of the four-year discounted cash flow for the two hypothetical investments under consideration is:

$$f(x) = \sigma^{-1}(2\pi)^{-\frac{1}{2}} \exp\left(-\tfrac{1}{2}\left(\tfrac{x-\mu}{\sigma}\right)^2\right)$$

By superimposing the density function of discounted cash flow on the utility function, the expected utility for the four-year discounted cash flow may be obtained as follows:

$$E\left[U(x)\right] = \int_{-\infty}^{\infty} U(x) \ f(x)dx$$

By completing the squares and then using transformations an expected utility for each proposed investment can be obtained (see Appendix).

A firm that follows the above procedure to determine the expected utility of four-year discounted cash flow for an investment thus has an additional criterion to determine whether to commit itself to the investment or not.

COMPARISON OF DECISION CRITERIA

Table 6-3 presents a comparison between the six objective decision criteria and the subjective decision criterion based on subjective utility for four-year discounted cash flow. Management can now consider all decision criteria, and decide accordingly.

This particular example shows that for six out of the seven decision criteria, proposal B is preferred over proposal A. However, a closer analysis reveals that the expected performance of the two proposals is quite similar for six of the seven criteria, and heavily in favor of proposal A for one criterion. Hence, management should probably decide in favor of proposal A.

137

Table 6-3

Comparison of Decision Criteria

| | Expected Value | | Best |
	Proposal A	Proposal B	Proposal
Internal rate of return	42.2%	42.5%	B
Baldwin rate of return	23.4%	24.9%	B
Non-discounted payout time in years	3.22	3.06	B
Discounted payout time in years	4.15	3.86	B
Present value for $r_o = .10$	$67.9 million	$42.7 million	A
Net discounted cash flow for 4 years for $r_o = .10$	$12.3 million	$12.4 million	B
Subjective utility based on net discounted cash flow for 4 years (0-1 scale)	.5991	.6018	B

Note: The utility functions parameters are: a = 4.44
b = .005
c = .085

EXERCISES

1. A research and development company is considering the
    purchase of a numerically controlled jig borer. If
    such a machine is purchased, four jig borer operators
    can be eliminated at an annual saving of $28,000.
    In addition, two tool makers can be eliminated at a
    savings of $16,500/year. The maintenance on the proposed
    machine is estimated to be $1,000/year. The maintenance
    on the five jig borers that would be replaced by the
    new machine totals $2,000/year. If the proposed machine
    is purchased, a programmer must be hired at a cost of
    $9,000/year. The installed cost of the new machine is
    $110,000. In addition, a controlled atmosphere is
    required and will cost $10,000. Operator training
    costs for the machine amounts to $5,000. The machine
    will have an economic life of 10 years with salvage
    value of $15,000.

    (a) Find the internal rate of return on the proposed
    machine.

    (b) Find the discounted and non-discounted payback in
    years if the cost of capital is 10 percent.

2. The Roanoke Machine Shop has four attractive investment
    opportunities consisting of machine modernization pro-
    grams. However, the available capital is insufficient
    to finance all four programs and a decision must be

made which one or ones to select.   The four investment
opportunities are shown below.

Alternative	First Year Investment	Cumulative Investment	Rate of Return in Percent
A	$30,000	$70,000	17.5
B	20,000	20,000	27.5
C	15,000	85,000	16.0
D	20,000	40,000	25.4

First year capital budget is $70,000.

Which projects would you choose with the given budget?
Will you make some other calculations before selecting
the project?

3.   Using the data for the two machines listed below, find

(a)   Rate of return,

(b)   Discounted present value,

(c)   Discounted rate of return,

(d)   Which one you will choose, and

(e)   What other information you may require before
investing.

Machine	A	B
Initial Investment	$32,000	$26,000
Forecasted life	7 yr	6 yr
Forecasted salvage value	$ 5,000	$ 2,000

Machine A

Year	Expenses	Revenue	Salvage value
1	$ 7,500	$18,000	$25,000
2	6,000	"	20,000
3	2,300	"	16,000
4	6,200	"	12,500
5	9,600	"	9,500
6	11,800	"	7,000
7	13,300	"	5,000

Machine B

Year	Expenses	Revenue	Salvage value
1	$11,000	$18,000	$22,000
2	8,000	"	16,000
3	3,000	"	11,000
4	6,000	"	7,000
5	9,000	"	4,000
6	12,000	"	2,000

4. Derive the expected utility $E[U(x)]$ if the utility function is,

$$U(x) = a[\exp(-bx - c)]$$

and the density function for discounted cash flow is the negative exponential function,

$$f(x) = \alpha \exp(-\beta x)$$

5. Discuss in an analytical way the difference between the regular internal rate of return and the Baldwin rate of return. Which rate would you recommend a firm should use for capital investment decision making?

6. Under which circumstances would you recommend that a firm not use expected monetary value for decision making?

7. Can the notion of utility be used without going to the trouble of estimating formal utility functions?

8. Consolidated Power and Light is faced with the decision to expand output by building a new thermal power plant or a gas turbine facility. The expansion is necessitated by expanded peak demand especially during summer heat waves. A thermal power plant will provide a permanent durable facility with a high installation, operation (except fuel) and capital cost. The gas turbine facility will have higher fuel cost but lower installation, operation and capital costs.

   The installation cost of the thermal power plant is $400 per kilowatt of capacity. Operation (except fuel) and capital costs (including depreciation) amount to a capitalized value of 10% of installation cost. Fuel cost amount to $.005 per kilowatt hour.

   The gas turbine installation will cost 100x percent less than the thermal power plant, but the turbine plant's fuel cost will be 100y percent higher. Find the relationship between x and y and the number of years h for which the discounted costs of the two types of equipment are identical.

9. Using the model developed in exercise 8 above and for values of x = 0.35, y = 1.25 and h = 25 find the number of hours of peak load operation at which Consolidated Power and Light would be indifferent between the thermal plant and the gas turbine plant.

10. The Park Edge Machine Company is considering two alternate machines for the purpose of performing a special operation on a gear blank. The investment, operating cost and life data for the two machines are shown below.

	Machine A	Machine B
Investment cost	$6000	$12000
Expected life in years	5	5
Fixed costs per year	$1100	$ 1600
Variable costs per piece	$1.55	$ 1.05

Maximum output on either machine is 20,000 pieces per year. However, demand is expected to range between 8000 and 12,000 pieces per year. Calculate internal rate of return for each machine for indicated demand range if the pieces are sold for $2 each. Also determine Baldwin rate of return for a capital cost rate of 10 percent.

11. Based on information in exercise 10 which machine is preferred at demand levels between 5000 and 20,000 pieces annually?

12. Park Edge Machine (see exercise 10) has a required discounted payback of two years. Would both machines satisfy that criterion? Show your work.

13. What levels of demand must be reached for Park Edge (see exercise 10) to attain a one year payback on each of the two machines?

14. Hoffman Machine and Tool is considering two pieces of machinery. Either machine is expected to generate annual revenues of $20,000. However, the acquisition cost, the operating cost of the two machines, and the salvage value after six years, the planning horizon, differ considerably. Machine A requires an initial investment of $35,000 and is expected to have a salvage value of $6000. Machine B costs $25,000 installed and is expected to have a salvage value of $3000. The annual operating expenses for each machine are as shown below.

Year	Machine A	Machine B
1	8000	12000
2	6500	8000
3	6500	7500
4	6000	6000
5	8000	7000
6	11000	9500

Find the payback, the discounted payback and the average annual earnings for each alternative.

15. If Hoffman Machine and Tool's (exercise 14) cost of capital is 10 percent per year for the next six years find the Baldwin rate of return, the internal rate of return and the present value for each alternative.

16. The student at this point is probably convinced that internal rate of return or the Baldwin rate of return are more desirable criteria to evaluate capital investments than payback or even discounted payback. With the above in mind, why do many firms still depend extensively and sometimes exclusively on simple payback?

APPENDIX

Utility Calculations

$$E\big[U(x)\big] = \int_{-\infty}^{\infty} a\big[1-\exp(-bx-c)\big]\, f(x)dx$$

$$= a\int_{-\infty}^{\infty} f(x)dx - a\int_{-\infty}^{\infty}\exp(-bx-c)\sigma^{-1}(2\pi)^{-\frac{1}{2}}\exp\Big[-\tfrac{1}{2}\Big(\tfrac{x-\mu}{\sigma}\Big)^2\Big]dx$$

$$= a - a\int_{-\infty}^{\infty}\sigma^{-1}(2\pi)^{-\frac{1}{2}}\exp\Big[-\tfrac{1}{2}\sigma^{-2}(x^2-2x\mu+\mu^2+2b\sigma^2 x+2c\sigma^2)\Big]dx$$

$$= a - a\,\exp(-\mu b + \sigma^2 b^2/2 - c)\int_{-\infty}^{\infty}\sigma^{-1}(2\pi)^{-\frac{1}{2}}\exp\Big[-\tfrac{1}{2}\Big(\tfrac{x-\mu+b\sigma^2}{\sigma}\Big)^2\Big]dx$$

Let $y = \dfrac{x-\mu+b\sigma^2}{\sigma}$, then

$$E\big[U(x)\big] = a\Big[1 - \exp(-\mu b + \sigma^2 b^2/2 - c)\Big]\int_{-\infty}^{\infty}(2\pi)^{-\frac{1}{2}}\exp(-y^2/2)dy\Big]$$

$$= a\Big[1 - \exp(-\mu b + \sigma^2 b^2/2 - c)\Big]$$

PART II.   CAPACITY PLANNING MODELS

Capacity planning is an activity that produces decisions which are usually irreversible in the sense that decisions made to engage in the building of a new production facility, the expansion of an existing facility or the change of an existing facility usually create conditions the firm must live with for a considerable period of time. Once a plant or other production facility is built, it is difficult to change it and the firm must be able to live with it. For the above reason capacity planning is so important.

This part will be concerned with analyzing capacity and input substitution possibilities prior to a production project coming to fruition. Of course analysis can also take place on any existing installation or configuration to determine how efficient the allocation of inputs is. The latter analysis may even indicate that, even though the fixed input such as buildings and/or equipment are already existing, savings may be realized by expanding or abandoning part of the fixed inputs.

The analysis in these next three chapters will evaluate combinations of stock inputs such as buildings and equipment and combinations of stock inputs and flow inputs such as labor, energy, etc. An example of a combination of several stock inputs is discussed in Chapter 8 where the capacity of a production facility can be reduced at the expense of the storage facility and vice versa. Chapter 7 concentrates on the theory and on simple examples of input substitution or rather on the technological level of a production facility. Several examples of stock and flow inputs such as the electrical and heat transmission problems are presented. Chapter 9 presents a case where capacity inputs of several parts of a production facility can be varied up and down in response to a specified output.

The objective of this part is to supplement the several systems analysis approaches discussed in the previous part. Chapter 7 reviews how many production processes can be framed in a production function framework. Once that is accomplished it then becomes straightforward to determine the minimum cost combination of inputs, be they stock or flow, given the output requirements. Chapters 8 and 9 both discuss more restricted, although more practical, applications of capacity and input substitution. Chapter 8 illustrates how a model was developed to plan the capacity of an oxygen plant for a basic oxygen steel making furnace. Chapter 9 reports on a study in a corrugated container plant of a large paper box manufacturer which investigated the optimal size of an intermediate product the firm was considering to produce.

CHAPTER 7.--CAPACITY AND INPUT SUBSTITUTION ANALYSIS

Capacity analysis is concerned with the substitution effects between current and capital inputs and between different kinds of capital inputs. This area of decision making is very critical for a firm because once the size, or capacity, of a capital input (investment) has been determined, the output capacity has been fixed for the short run, and in many cases also for the long run.

INPUT SUBSTITUTION ANALYSIS

The major characteristics of most durable capital inputs are 1) their presence is required if production is to occur, 2) the degree of consumption or wear, measured by depreciation, of such inputs is insignificant over the very short run, and 3) the capacity is essentially continuously variable prior to installation. However, once the capacity decision has been made and put into effect, changes and especially reductions in capacity cannot be made.

Therefore, a proper framework in which capacity decisions can be made is extremely important. To this problem area this and the next two chapters are directed. In this chapter the treatment will center on the more theoretical approach which is not without potentially real applications as will be revealed in the examples. In the next two chapters two models will be presented which emanated directly

from real life problems and were solved by analytic techniques.

The bulk of this chapter is based on work done by
Smith[1] and Chenery.[2] The material is also related on a more
practical basis to work done by Manne[3] and Terborgh.[4]
Most of the analysis is based on the hypothesis that many
production processes can be explained by a production function defined over a class of imperfectly substitutable
inputs. Among these inputs are both current inputs such
as labor, raw material, and power, and the durable inputs
required for production. A production function, which
relates the rate of production output to the rate of current
input such as labor, etc. and the durable input level, such
as machinery, etc. provides an important connecting link
between short and long term decisions over the firm's planning horizon. To incorporate the above concepts extensive
use of the stock-flow production function will be made.

---

[1]V.L. Smith, Investment and Production, Cambridge,
Massachusetts: Harvard University Press, 1961.

[2]H.B. Chenery, "Engineering Production Functions,"
Quarterly Journal of Economics, Vol. 63, November 1949,
pp. 507-31.

[3]A.S. Manne, Investments for Capacity Expansion, Cambridge, Massachusetts: The M.I.T. Press, 1967.

[4]G. Terborgh, Dynamic Equipment Policy, New York:
McGraw-Hill, 1949.

The term stock refers to the capacity of the production facility in terms of size of durable input, and the term flow refers to the input flow rate such as labor cost per week, etc.

A typical example of the stock-flow production is the case of the electrical transmission cable. A mechanism of substitution is present in the electrical power transmission process. This power transmission process is really a production process in its own right. For a required power output rate a larger diameter cable, i.e. a larger capital investment, will decrease the resistance to current flow and less power will be consumed. In other words the investment in the stock, the transmission cable, can be reduced at the expense of the flow, the power input, and vice versa.

Once it has been decided to use production functions for analysis there arises the problem of how to determine the production function. This is a relatively easy task in pure theoretical work where one just hypothesizes a specific production function such as, for instance, the Cobb-Douglas function.[5] Where experimentation is possible it is also relatively easy to determine a production function.

---

[5] The Cobb-Douglas production function is homogeneous of degree one and has the form $y = Ax_1^{\alpha} x_2^{1-\alpha}$. It was originally proposed by C.W. Cobb and P.H. Douglas, "A Theory of Production," American Economic Review, Supplement, March 1928, pp. 139-65.

Heady[6] principally used the method of experimentation to
analyze and develop agricultural production functions. Both
Tintner[7] and Nicholls[8] have used the experimental method to
derive production functions. Also, in most engineering processes the technical principles involved in relating inputs
to each other and to output are sufficiently simple to per-
mit their manageable analytical representation. However, in
many other production processes it is very nearly impossible
to derive meaningful production functions, although the
advent of the computer with its attendant simulation possibilities may generate an increase in the experimental
methods for deriving production functions.

MATHEMATICAL PROGRAMMING FRAMEWORK

Before continuing with capacity analysis, the analogy
of the capacity analysis, to be presented in this chapter,
and the more familiar linear programming framework will be
presented. In a typical linear programming cost minimization

---

[6]E.O. Heady, "An Econometric Investigation of the Technology of Agricultural Production Functions, Econometrica,
Vol. 25, April 1957, pp. 249-68.

[7]G. Tintner, "A Note on the Derivation of Production Functions from Farm Records Data," Econometrica, Vol. 12, No. 1
(January 1944), pp. 26-34.

[8]W.H. Nicholls, Labor Productivity Functions in Meat
Packing, Chicago: Chicago University Press, 1948, p. 12.

model, an objective function is minimized subject to a set of constraints usually consisting of output parameters.

Linear Programming Model

A typical linear programming cost minimization model is written as,

$$\text{Minimize } C = \sum_{i=1}^{n} w_i x_i,$$

subjective to

$$\sum_{i=1}^{n} u_{ij} x_i \geq \bar{y}_j \quad J = 1, 2, \ldots, m$$

where the x_i's are the n input decision variables, the w_i's are the cost coefficients, the u_{ij}'s are the substitution coefficients in the constraint matrix, and the $\bar{y}_j$'s are the minimum output requirements of the m outputs.

Analogous to the above linear programming framework is the capacity analysis model. The capacity analysis model, unfortunately, has no linear substitution relationships, and therefore is considerably more difficult to solve than the well-developed solution techniques available for a problem formulated by the linear programming framework.

Capacity Analysis Model

The general formulation of a capacity analysis model consists of

$$\text{Minimize } C = f(x, z; w),$$

subject to
$$g_j(x,z;u_j) \geq \bar{y}_j \quad j = 1,2,\ldots, m$$

$$x,z \geq o \qquad\qquad (7\text{-}1)$$

where x and z are decision vectors, and w and u_j are parameter vectors. In the examples to follow x and z are actually decision variables or one-dimensional vectors. The decision vector x includes all current flow variables and the vector u the capital stock variables. The constraints, $\bar{y}_j$'s, are the output requirements. Typically, there is only one output requirement. However, if more than one product is produced, each having its own unique production function, then there would be an equal number of output requirements.

Examining the linear programming model and the capacity model reveals that both belong to the same general model. The linear programming formulation, usually considered a general model because of its wide acceptance and extensive application, is really just a special application of the general model presented by (7-1).

In the next chapter a slightly different model will be investigated. The model is an extension of the general mathematical programming model discussed above and includes a number of special features.

FORESTRY PROBLEM

This chapter will be mainly concerned with production functions derived from technical principles or empirically-based relationships. However, the first example will illustrate the methodology of production function derivation for the somewhat hypothetical case of tree production.[9]

Tree Production

In producing or growing trees a growth function, $f(t)$ is postulated. This function gives the number of cubic feet of usable lumber or pulp in a tree as a function of its age, t, in years. It is assumed that the function represents the basic technological data of the growth process and comes from the principles of arboreal ecology. It is also assumed that the annual required output rate, y, is constant and consists of mature trees cut from a forest with an even-age distribution of trees. The input consists of saplings which are planted at an even annual rate, x. Based on the assumptions for the static process described above the output of timber per year is,

$$y = xf(t). \qquad (7-2)$$

[9] The tree production example is based on F. Lutz and V. Lutz. The Theory of Investment of the Firm, Princeton: Princeton University Press, 1951, and on V.L. Smith, op.cit., p. 19.

The above process may also be viewed as a black box with a uniform input, a uniform output, and a certain amount of goods-in-process.

Expression (7-2) is a sort of production function in that it relates output to input. However, it is not a stock-flow production function because stocks of capital, land, do not appear in it. This can be corrected by the introduction of the variable, z, for land requirements. If we measure z in tree-units, that is the amount of land required to grow a single tree, then

$$z = xt$$

because xt is the total planting over any t-year period. If the parameter t is eliminated from (7-2) the following stock-flow production function is obtained,

$$y = x f(z/x) = g(x,z).$$

Suppose that the growth function has the mathematical form

$$f(t) = t^{\alpha} = x^{-\alpha} z^{\alpha}, \qquad o < \alpha < 1$$

then

$$y = x^{1-\alpha} z^{\alpha}, \qquad\qquad (7-3)$$

which is the Cobb-Douglas production function with constant returns to scale.

Input Substitution

In the above process substitution between the current input, saplings, and the capital input, land, occurs because of the possibility of varying the growth period of trees. A constant output can be produced with less land if the growth period is shortened which requires an increased input of saplings.

The above formulation has ignored one additional investment, the investment of inventory-in-process. In order to produce the synchronized state of the process, t years of planting x saplings per year are necessary. The inventory of partially grown trees at any time, t, is then

$$u(t) = \int_{0}^{t} x f(v) dv$$

or for $f(t) = t^{\alpha}$,

$$u = \frac{xt^{\alpha+1}}{\alpha+1} = \frac{x^{-\alpha}z^{\alpha+1}}{\alpha+1}$$

which is the inventory-in-process in terms of x and z.

In the context of the tree problem the primary objective of determining capacity means determining the amount of land required for a constant rate of output. Assuming that the firm's objective is to minimize cost on current account, the hypothetical cost function shown below is proposed,

$$C = w_1 x + w_2 z + w_3 u$$

$$= w_1 x + w_2 z + w_3 x^{-\alpha} z^{\alpha+1}/(\alpha+1) \qquad (7\text{-}4)$$

The coefficient w_1 is the cost per sapling, w_2 and w_3 may be thought of as the cost per year of maintaining the presence of a unit of capital in production. The objective is therefore to minimize (7-4) subject to the constraint (7-3).

ELECTRICAL ENERGY TRANSMISSION

This problem is very common. Electrical energy in the form of alternating current is to be moved from one location, the generating source, to another location, the consumption outlet. This is a simplified version of the real problem, but simplification will allow the problem to be analyzed more thoroughly. The two major substitutable inputs are transmission cable, the stock, and electrical energy, the current input flow. For a constant distance over which electrical energy is to be transported, variation in the quantity of cable input can only occur because of the possibility of choosing among different cable sizes. Hence, the cable could be measured in cross-sectional area. The energy input, as well as the required energy output at the end of the transmission line could be measured in kilowatt-hours.

Input Substitution

To obtain a relationship between the current input, kilowatt-hours, and the capital input, cable, several well-established engineering laws will be used. A well-known engineering phenomenon is that the greater the resistance of the wire the greater the energy loss from heat generated in the wire. Hence, smaller cable sizes require larger energy inputs for equivalent output requirements. Therefore, the substitution is between cable size and energy input rate.

Four empirical laws of physical science, the law of energy conservation, the law of heat dissipation in electrical circuits, Ohm's law, and a law concerning the conductor and resistive properties of materials, enables one to derive the required production function for the electrical transmission process.

Law of Conservation of Energy

The law of the conservation of energy for any electrical system requires that

$$P_O = P_I - P_L \qquad (7-5)$$

where P_O is the required power output rate of the transmission line, P_I is the power input at the input source, and P_L is the power loss in the transmission line due to resistance.

Law of Heat Dissipation

According to the law of heat dissipation, the energy lost in the form of heat in a conductor of electricity is equal to the square of the electrical current flowing through the conductor times the electrical resistance of the conductor. Stated in symbolic terms it becomes,

$$P_L = I^2 R$$

where I is current flow and R is electrical resistance of the transmission line in ohms.

Ohm's Law

If E_O is the effective or "root-mean-square" voltage at which customers require energy to be delivered, and if $\cos \phi$ is the power factor of the customer's load,[10] then from Ohm's law,

$$P_O = I E_O \cos \phi$$

or

$$I = P_O/(E_O \cos \phi)$$

---

[10] It was noted in Chapter 4 that the power factor can be adjusted to an optimal level within a plant. That is on a micro basis the power factor is adjustable. The power supplier, however, has little control over the resultant value of the power factor except by penalty charges for excessive deviations of acceptable levels.

Resistivity Law

Finally, using the empirically derived and known
coefficient of resistivity, ρ, the cross-sectional area of
the cable, A, and the return-trip length of the power line,
L, the resistance R of the transmission line can be derived
as follows,

$$R = \rho L/A$$

Substituting back into the energy conservation relationship (7-5) the following results are obtained,

$$P_o = P_I - P_o^2/(E_o \cos \phi)^2 \, \rho L/A \qquad (7\text{-}6)$$

Transforming the engineering variables in (7-6) into
the more universal economic variables the output rate P_o
becomes y, the input rate P_I becomes x and the capital stock
required in the form of pounds of cable dLA becomes z where
d is the density in pounds of cable per cubic foot. Using
the economic variables the production function (7-6) can
now be expressed as,

$$F(y,x,z) = z(x\text{-}y) - ky^2 = 0 \qquad (7\text{-}7)$$

where $k = L^2 y^2 \rho d/(E_o^2 \cos^2 \phi)$.

Constraints on Input Variables

The above derivation, of course, assumes that there
are no constraints on the input variables x and z. In

practice, one can expect that the power input rate x is definitely governed by the power source, i.e., the generating source. Although x can be varied considerably one would expect that certain discrete values are more economically desirable than a continuous range of values. Similarly, the cable input z will be restrained from above and below. Too small a cable size may be too weak to support itself, and too large a cable size may produce overheating. In addition, for very high voltages the corona effect[11] may introduce an additional power loss.

To select the optimum combination of power input rate and cable size it will again be assumed that the objective is to minimize cost on current account. Assuming that the transmission line will have a lifetime of T years, the objective function to be minimized becomes,

$$C = w_1 x + w_2 z \qquad (7\text{-}8)$$

where w_1 is the power cost of the current input and w_2 is annualized cost of keeping one unit weight of the cable in operation. The latter item presumably would include capital cost (interest) as well as depreciation. The objective function (7-8) will then be minimized subject to (7-7) for a

---

[11]When the voltage between two conductors is very high the electric field, which is greatest at the surface, ionizes the air at the surfaces of the conductor. The ionization of the air greatly increases its conductivity and substantial power losses result.

specified energy output of $\bar{y}$.

HEAT TRANSMISSION

Another engineering problem which can be transformed neatly into a production function framework is the problem of determining the amount of insulation to use on a heat transmission device.[12] The problem is concerned with the mechanism of substitution between a current input, heat, and a capital input, amount of insulation.

Conservation of Energy

Consider heat input through the medium of steam or hot water from a boiler or powerhouse is H_I. Heat loss, H_L, from the law of the conservation of energy is then,

$$H_L = H_I - H_o$$

and

$$H_o = H_I - H_L \tag{7-9}$$

Formula (7-9) is analogous to formula (7-5) in the electrical transmission problem.

---

[12]This problem has been adapted from one shown by H.E. Schweyer, _Process Engineering Economics_, New York: McGraw-Hill, 1955, pp. 211-13 and from V.L. Smith, op.cit., pp. 30-37.

Conductivity Relationships

The relationship between H_L and the thickness, t, of insulation used is given by the empirical approximation formula,

$$H_L = \frac{T_I - T_o}{K + t/kS}$$

where T_o is the average air temperature outside the pipe, T_I is the temperature of the steam in the pipe, and K is a constant depending upon the thickness and thermal conductivity properties of the pipe, the steam film just inside the pipe, and the air film layer just outside the pipe. The constant k is the thermal conductivity of the insulating material of thickness, t, and S is the surface area of the pipe which is equal to $2\Pi rL$.

Rewriting (7-9) results in,

$$H_o = H_I - \frac{T_o - T_I}{K + t/kS} \qquad (7\text{-}10)$$

Changing the engineering variables in (7-10) to economic variables that is, letting $y = H_o$, $x = H_I$, and $z = St$, results in,

$$y = f(x,z) = x - \frac{T_o - T_I}{K + z/kS}_2 \qquad (7\text{-}11)$$

To minimize cost on current account can be accomplished with the objective function,

$$C = w_1 x + w_2 z$$

subject to (7-11) for a specified output $\bar{y}$.

MACHINING PROBLEM

In most machining problems cost is used as the criterion for determining the best engineering method or design. Problems of this sort can usually be solved by the production function approach and the minimization of the cost of the current input and the current cost of the capital input.

The Substitution Process

The life of cutting tools in metal-cutting processes depends on the cutting speed and depth of cut, or more realistically on the metal-removal rate. For a higher metal-removal rate fewer machines are required than for a lower metal-removal rate. Hence, substitution occurs between cutting-tools, the current input, and machines, the capital input.

Production Function Formulation

A production function will be formulated for the metal-removal process which produces cylindrical bars of metal. These bars are produced by turning on a lathe. Let V be the cutting speed of a machine in lineal feet per hour along the bar being machined. Let T be the life of a

cutting tool in machine hours, then the tool life and cutting
speed can be related by the empirical engineering formula,

$$VT^\alpha = \beta \tag{7-12}$$

where α and β are constants depending upon the depth of cut,
type of metal, and other considerations.

If z stands for the number of lathes required, and
x for the consumption rate of cutting tools, then

$$x = z/T \tag{7-13}$$

because the dimension for z is [lathes] and if T is [lathe
hours/tools] then the x dimension is [tools/hour].

Assume that the depth of cut is fixed and the total
number of lineal feet of metal that must be removed by
machining to produce a finished piece is m, then m/V is the
total number of machine hours required per piece. Using
z lathes, the output per hour of finished pieces, y, is
equal to z divided by the ratio m/V or,

$$y = zV/m \tag{7-14}$$

Combining the three relationships in (7-12), (7-13),
and (7-14) the production function can be expressed explicitly
in terms of x and z as follows,

$$y = kx^\alpha z^{1-\alpha} \tag{7-15}$$

where $k = \beta/m$.

Cost Model

It is assumed that the objective is to minimize cost on current account. A current cost is associated with the current input, tools, and an annualized current cost is associated with the capital input, lathes. The annualized cost of the capital input consists of a capital or interest cost and a depreciation cost; both of these cost components are included in the cost coefficient w_2. Assuming a linear cost function the objective function becomes again,

$$C = w_1 x + w_2 z \qquad (7\text{-}16)$$

which is minimized with respect to (7-15) for a given output $\bar{y}$.

Example of Machining Problem

Suppose that the value of α is 0.25 which is obtained from empirically-determined tabulated data. The total amount of lineal feet to be removed from each piece, m, is 4. The life of a cutting tool in machine hours, T, is 16. The cutting speed, V, is 200 lineal feet per hour. From formula (7-12) it is found that $\beta = 400$, and the following production function is obtained,

$$y = 100x^{.25}z^{.75}$$

If it is assumed that the cost function is of the general linear form shown in (7-16) with w_1 equal to 5 and w_2 equal to 10, then the problem can be solved by the Lagrangean multiplier technique. The Lagrangean form is shown below,

$$L = 5x+10z-\lambda[y-100x^{.25}z^{.75}]$$

Taking partial derivatives with respect to x,z, and λ provides the three equations,

$$\frac{\partial L}{\partial x} = 5 + .25x^{-.75}z^{.75}\lambda$$

$$\frac{\partial L}{\partial z} = 10 + .75x^{.25}z^{-.25}\lambda$$

$$\frac{\partial L}{\partial \lambda} = 100x^{.25}z^{.75} - y$$

Setting the three equations equal to zero provides three equations which when solved simultaneously provide the minimum cost solution. Unfortunately, x, z, and λ cannot be solved for directly. However, a simpler relationship can be quickly derived which reveals that the ratio of input x to input z is 2 to 3. Therefore, for a given output of $\bar{y}$ is 1000, x and z can be found.

STAFFING PROBLEM

The maintenance facility staffing model solves the problem of determining how large the maintenance staff should be such that the total cost of maintenance staff and equipment is minimized. The implicit assumption underlying the model is that maintenance manpower can be substituted for equipment and vice versa. The maintenance staff causes the incurrence of a current flow cost and acquisition of equipment causes the incurrence of a capital stock cost.

A Queuing Theory Problem

Using the well-developed theory and assumptions of queuing theory the mean arrival rate per unit time for equipment to be serviced will be represented by λ. It will be assumed that the arrival rate is distributed according to the Poisson distribution. The time required to service a piece of equipment is assumed to be distributed according to the negative exponential distribution with mean μ. Hence μ is the number of pieces of equipment serviced per unit time.

Let n be the number of units in the queue, including the one being serviced. Also let $P_n(t)$ be the probability of n units in the queue at time t. Then, if $P_n(t) = P_n$, it follows that,

$$P_o = 1 - \lambda/\mu$$

and the mean n is,

$$\bar{n} = \lambda/(\mu-\lambda) \qquad\qquad (7\text{-}17)$$

Assuming that servicing or maintenance needs are proportional to the firm's output, y, in operating hours, miles, etc., results in,

$$\lambda = \alpha y \qquad (7\text{-}18)$$

where α is the maintenance arrival rate per unit of output.

Also assume that the maintenance rate is proportional to the size of the maintenance staff, x, and can be expressed as,

$$\mu = \beta x \qquad (7\text{-}19)$$

where β is serviced pieces of equipment per maintenance staff member.

If the average number of output units a piece of equipment produces between maintenance is m, then

$$y = m\bar{N} \qquad (7\text{-}20)$$

where $\bar{N}$ is the mean number of pieces of equipment in operation during a unit time period.

Production Function Derivation

The total stock of equipment required consists of those units **waiting for or in** maintenance plus the units in operation. Hence

$$z = \bar{n} + \bar{N} \qquad (7\text{-}21)$$

Formula (7-21) can be changed to,

$$\bar{N} = z - \bar{n} \tag{7-22}$$

and from (7-20) and (7-22) one can derive,

$$y = mz - m\bar{n} \tag{7-23}$$

From (7-17), (7-18) and (7-19) one can similarly derive,

$$\bar{n} = \alpha y/(\beta x - \alpha y) \tag{7-24}$$

Combining (7-23) and (7-24) results in,

$$y = mz - m\alpha y/(\beta x - \alpha y) \tag{7-25}$$

Since y is found on either side of the equal sign, and further simplification is not possible the production function cannot be stated in explicit form. However, in implicit form the production function becomes,

$$F(y,x,z) = y + m\alpha y/(\beta x - \alpha y) - mz = 0 \tag{7-26}$$

Equation (7-26) can also be transformed to the quadratic form,

$$y^2 - (mz + m + \beta x/\alpha)y - m\beta zx/\alpha = 0 \tag{7-27}$$

Applying the capacity model discussed before the aim is now to minimize cost on current account. Hence,

$$\text{Min } C = w_1 x + w_2 z$$

subject to (7-27) for a given or minimum output $\bar{y}$.

Staffing Example for a Maintenance Facility

Suppose Mercury Airlines would like to determine the optimum staffing level of its maintenance base for its twin engine Super Jets currently on order. Because of the high degree of education and training required the staff of the maintenance facility will remain fixed once the staffing level has been determined.

Mercury has a total of 45 twin engine Super Jets on order and each one will require maintenance on the average every 200 operating hours.

Total required operating hours of the fleet ($\bar{y}$) amount to 20000 hours per month. Hence, the arrival rate of airplanes at the maintenance base (λ) will amount to 100 airplanes per month.

From formula (7-18) the symbol α can be determined which has the value of 0.005 and stands for the number of times an airplane requires maintenance per operating hour.

Each time a Super Jet is brought into the maintenance base it takes the equivalent of ten man-months to service the aircraft. Hence, the number of serviced planes per man-month (β) equals 0.1. The number of hours an aircraft could operate per month if it required no service (m) equals 500 hours. Since each aircraft requires servicing every 200

operating hours the actual number of hours per month an aircraft operates is, of course, considerably less.

All required parameters for the implicit production
function (7-26) have been specified and it is possible to
solve for a set of feasible pairs of values for (x,z).
Table 7-1 lists the various combinations.

Table 7-1

Feasible Alternatives for Staffing Problem

Flight hours per month (y)	Number of airplanes (z)	Employment level of maintenance plant (x)
20000	41	2000
	42	1500
	43	1334
	44	1250
	45	1200
	46	1167
	47	1143
	48	1125
	49	1112
	50	1100
	60	1050

Based on formula (7-20) the mean number of airplanes
in operation, $\bar{N}$, equals forty and with the minimum number
of airplanes, 41, in operation as shown in Table 7-1 only
one airplane could be out of service on the average.

Figure 7-1 shows an isoproduct map of the employment
level and the number of airplanes in the fleet. Based on

174

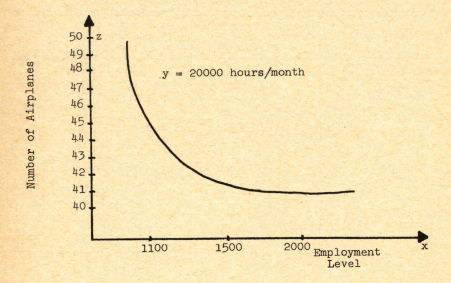

Figure 7-1

Isoproduct Map of Staffing Problem

the cost of Super Jets and the personnel and maintenance
facility costs the optimal level of both the size of the
fleet and the maintenance facility can be determined.
The cost formula can be stated as,

$$C = w_1 x + w_2 z.$$

EXERCISES

1. Find the production function, inventory of partially-
   grown trees at time t, and aggregate cost function for
   the tree production function if the growth function is
   $f(t) = \alpha t$. Repeat for $f(t) = \alpha_0 + \alpha_1 t - \alpha_2 t^2$.

2. Plot the production function of the electrical energy
    transmission problem with cable input (in lbs.) on the
    y-axis and electrical energy input (in kilowatts) on
    the x-axis. Use the following parameter values. Ohms
    resistance per pound mile of cable, $\rho dL = 900$, $L = 200$
    miles, $E_o = 60000$ volts and $\cos \phi = 0.75$. The curves
    are to be plotted for output (y) values of 1000, 2000
    and 3000 kilowatts.

3. Plot the production function of the heat transmission
    problem with insulation (in cubic feet) on the y-axis
    and heat input (in thousands of BTU per hour) on the
    x-axis. Use the following parameter values. The tem-
    perature of saturated steam in the pipe is 360° F, the
    ambient room temperature is 80° F, K is 0.014, k is
    0.04 BTU per hour per square foot per inch, and S is
    250 square feet. The curves are to be plotted for
    output (y) of 5000, 10000 and 15000 BTU per hour.

4. In the heat transmission problem the amount of insula-
    tion, the stock input z, is set equal to St where S is
    the surface area of the pipe which is equal to $2\Pi rL$, r
    being the radius of the pipe and L the length, and t is
    the thickness of the insulating material. However,
    setting z equal to St is not a good approximation
    especially for heavy thickness of insulation. Develop

a better approximation for z.

5. Based on the previous problem, assume z is approximated
    by $2S(R-r)/3$ where R is the radius of the pipe including
    insulation and r is the radius of the pipe without
    insulation. Develop the production function.

6. Suppose the production function for the machining problem
    is $y = 75 x^{.33} z^{.67}$, and the cost function is $C = 6x+9z$.
    Solve for x and z.

7. Plot the production function for the staffing problem
    of a truck maintenance facility. The expected number
    of trucks should appear on the y-axis and the number of
    man hours of labor per year on the x-axis. The para-
    meters values are: $m = 100,000$, $\alpha = 1$ and $\beta = 100$.
    The curves are to be plotted for output (y) values of
    one million, two million and three million truck miles
    per year.

8. If the cost of keeping an additional airplane, over and
    above the minimum level of 41 airplanes, in operation
    amounts to $75,000 per month, and the cost of a staffed
    maintenance facility amounts to $2,000 per month per man
    find the optimum number of airplanes and the staffing
    level of maintenance facility.

9. The production function for corn production is of the following form,

$$y = \beta x_1^{\alpha_1} x_2^{\alpha_2} x_3^{1-\alpha_1-\alpha_2}$$

where x_1, x_2, and x_3 are inputs such as land, fertilizer and labor, y is output in pounds or bushels of corn and α_1, α_2 and β are coefficients, $0 \le \alpha_1 \le 1$, $0 \le \alpha_2 \le 1$. The cost function is of the form

$$C = \alpha_0 + \alpha_1 x_1 + \alpha_2 x_2 + \alpha_3 x_3$$

For a given amount of land and a specified output how would you select optimum values of fertilizer input and labor input?

10. The production function for electricity of a steam generating power plant is of the following form,

$$y = \exp(a + bx + cz)$$

where $a, b, c > 0$; y is output in kilowatt-hours, x is the input of coal in tons per day and z is the stock input in kilowatt capacity units. Develop the cost minimization model for electricity production and plot the indifference curve between x and z for a fixed output of y.

APPENDIX

All the cost functions and production functions discussed in this chapter deal with costs and output flow rates per unit of time. Input, on the one hand, consists of one or more current or flow inputs plus one or more capital or stock inputs. If a minimum cost combination of current input and a capital input is to be obtained the current input must be converted to an equivalent capital input or the capital input must be converted to an equivalent current input. The latter approach has been taken and will be discussed here.

The easiest case arises when the capital stock is of infinite duration, such as land, and has never to be replaced. In that case a current cost w_2 can be assigned to the capital stock equal to the capital or interest cost of the firm. Then, if r is the rate of interest and z is the total cost of the capital stock, w_2 is equal to r.

Unfortunately, few productive investments in capital stock have an infinite life. It is much more common to encounter capital stock investments which have a definite or approximate life span. The life span will be identified by the symbol L. Then the capital cost is the interest cost of the initial capital stock z, which shall be called z´, plus the replacement cost of the capital stock z´ every L time periods or say years. If the original investment of z and all subsequent investments of z´ could be combined into one common value, say a present value, then an interest cost could be assigned to this present value of the initial and all future investments. This will be accomplished in the next paragraph.

The first capital stock outlay consists of the initial investment z'. L periods later, an additional investment of z' is required but this investment can be discounted to the present by the discount factor $1/(1+i)^L$. Similarly, 2L periods later, an additional investment of z' is required which is discounted at $1/(1+i)^{2L}$, and so forth until theoretically infinity is reached. The total of all capital stock investments can thus be summarized as follows

$$z' + z'/(1+i)^L + z'/(1+i)^{2L} + \ldots$$

which is a simple geometric progression with a constant ratio of $1/(1+i)^L$, and can be summed to the total present value as follows

$$z = z'/[1-(1+i)^{-L}] \qquad\qquad (A7.1)$$

Just multiplying z by the interest cost i for which the symbol w_2 is used produces a cost function in terms of current cost only.

It is possible and convenient to rewrite formula (A7.1) in continuous form. Before this is done the relationship between the discrete and continuous discount factors will be discussed.

A temporary digression from the discount factor discussed above is required to allow a brief discussion of the compounding of interest which in a sense is the reverse of compounding discounts. If an amount v_o is invested at a compound rate of 100 i percent per year, then after n years the accumulated total will be V_n, where

$$V_n = v_o(1+i)^n$$

Suppose that instead of compounding interest at the end of every year, it is compounded k times during the year. In that case

$$V_n = v_o(1+i/k)^{nk}. \qquad (A7.2)$$

Formula (A7.2) can also be written as,

$$V_n = v_o[(1+i/k)^{k/i}]^{in}.$$

Introducing p = k/i, results in

$$V_n = v_o[(1+1/p)^p]^{in} \qquad (A7.3)$$

Formula (A7.3) can now be simplified as follows. The base of the natural logarithm, denoted by the letter e, is defined by the limit

$$e = \lim_{p \to \infty} (1+1/p)^k$$

Hence, the limit of formula (A7.3), as $p \to \infty$ and, therefore, k→∞ can be written as

$$V = \lim_{p \to \infty} V = \lim_{p \to \infty} [(1+1/p)^k]^{in} = v_o e^{in}$$

Returning now to formula (A7.1) note that it can be written as

$$z = z'/(1-e^{-it})$$

where L has been replaced by t which is a more customary symbol for continous interest formulas.

CHAPTER 8.--CAPACITY PLANNING-BASED ON EXOGENOUS RELATIONSHIPS[1]

In the previous chapter substitution possibilities between current flow inputs and capital stock inputs were discussed. Exogenous engineering relationships were used to derive analytical production functions which were then utilized in the cost minimization model.

In this chapter substitution possibilities again provide the opportunity to develop a cost minimization model. However, instead of analyzing stock-flow substitution possibilities, analysis will be concentrated on substitution possibilities in the capacities (sizes) of various production facilities. The proposed model will determine the capacities of the various production facilities so that total cost of manufacturing over some specified time horizon is minimized. Again, as in the previous chapter, exogenous engineering relationships will be used to derive what are called capacity functions of the various production facilities.

Although the work done in the area of optimal combinations of capacities is limited, several references should be noted. Smith[2] as was extensively discussed in the previous chapter contributed to the general theory of production by attacking

---

[1]This chapter is based on F.C. Jen, C.C. Pegels and J.M. Dupuis: "Optimal Capacities of Production Facilities," Management Science, Vol. 14, No. 10 (June 1968), pp. B573-B580.

[2]V.L. Smith, Investment and Production, Cambridge, Massachusetts: Harvard University Press, 1961.

the problem of production function formulation and production

system and multi-system optimization, without, however, solving

specifically the problem of optimal combinations. Aris[3],

Roberts[4], and Mitten and Nemhauser[5,6], on the other hand, ap-

plied the dynamic programming technique to study the chemical

reaction processes by converting a multistage optimization

problem into a series of one-stage optimization problems. The

technique cannot, however, be applied to problems where the

relationships between the production facilities in respect to

various decision variables are complex and interrelated. This

chapter presents a model that can be used to determine the

optimal (minimum cost or maximum profit) combination of the

capacities of the production facilities of a steady-state pro-

duction system when both the cost functions and the relation-

ships between the decision variables are complex. A specific

model, based on the general model, is formulated to solve the

practical problem of designing an oxygen producing system.

---

[3]R. Aris, The Optimal Design of Chemical Reactors, New York: Academic Press, 1961

[4]S.M. Roberts, Dynamic Programming in Chemical Engineering and Process Control, New York: Academic Press, 1964.

[5]L.G. Mitten and G.L. Nemhauser, "Multistage Optimization," Chemical Engineering Progress, Vol. 59, No. 1 (January 1963), pp. 52-60.

[6]L.G. Mitten and G.L. Nemhauser, "Optimization of Multi-stage Separation Processes by Dynamic Programming," The Canadian Journal of Chemical Engineering, October 1963.

GENERAL MODEL

The model presented below can be used to determine the optimal capacities of the production facilities used in a production system (plant) employing a steady-state continuous process. A particular feature of the problem for which this model is designed is that the demand for the product is not only deterministic but also cyclic or repetitive within such a short period of time that it precludes the adjustment of the level of production of the plant. Because of these particular features, the objective of the model can be stated as the minimization of the current cost (C) of producing a given output over the cycle, where current cost is defined to include all current outlays plus depreciation. Mathematically, the model can be stated as:

$$\text{Minimize } C = f(y_1, y_2, \ldots, y_n)$$

$$\text{Subject to: } A_i \leq X_i \leq B_i \text{ for } i = 1, 2, \ldots, n$$

$$\text{Where: } y_1 = g_1(X_1; Z_1)$$

$$y_2 = g_2(X_2; Z_2)$$

$$\vdots \quad \vdots$$

$$y_n = g_n(X_n; Z_n)$$

$$X_i \subset (x_1, x_2, \ldots, x_m) \quad i = 1, 2, \ldots, n$$

$$Z_i \subset (z_1, z_2, \ldots, z_p) \quad i = 1, 2, \ldots, n$$

The variables y_i, $i = 1, \ldots, n$ are variables denoting the capacity of the n production facilities. The vectors Z_i are parameter vectors, or vectors of known constants such as demand levels, unit costs of input, etc., which, together with the decision variable vectors X_i, determine the optimal capacities of the n machines by means of the relationships specified in the functions g_i.

Solution Methods

Two approaches can be tried to solve for the optimal values of the decision variables. Classical calculus techniques, including the use of Lagrangean multipliers are highly desirable. However, they are only satisfactory for functions in simple forms, e.g., quadratic forms. With complicated functions, the optimal values of the decision variables will be difficult to solve in most cases and it will be at best arduous to check the secondary conditions.[7] For example, the problem to be discussed later cannot be solved through classical calculus techniques.

Another approach is the use of a gradient method.[8] With the aid of a computer, the method can be quite efficient if only a few decision variables are involved and if the numerical

---

[7]Hessian matrices, of course, are the secondary conditions for the multivariate case. Note that the use of the Hessian matrices can only insure that the stationary point reached is a local extremum.

[8]For a brief explanation of the advantages and disadvantages of the gradient method see C.R. Carr and C.W. Howe: Quantitative Decision Procedures in Management and Economics, New York: McGraw-Hill, 1964, pp. 264-67. The gradient method is also described near the end of this chapter.

values of the parameters of the system are known.   However, as is well known, the gradient method does not insure the solution to be the global extremum.   The limitation is nevertheless not unique for the gradient method because it is applicable to the classical calculus techniques also.

APPLICATION TO AN OXYGEN PRODUCTION SYSTEM

We will now illustrate the use of the general model developed above to the specific problem of determining the optimum capacity of the production facilities that combine to make an oxygen producing system.

The Problem

Oxygen for the Basic Oxygen Furnace (BOF) is produced by a highly automated production facility. The demand for the oxygen is repetitive or cyclic approximately every hour, although the plant has to have a fixed constant output because the cycle is too short to allow the producer to adjust the level of production. Thus, a producer can build a plant that can produce at the maximum demand level and vent the excess production to the atmosphere during low-demand periods. Alternatively, the producer can build a small plant, compress the excess production during the low-demand period by a compressor and then store it in an inventory storage unit for use during the high-demand period.[9] Which alternative a producer should use is the problem the proposed model will solve.

The Model

In the context of the theoretical model developed in the preceding section, the problem can be regarded as finding the optimal capacities of three machines, i.e., an oxygen production machine, an inventory processing unit (a compressor) and an inventory storage unit, that combine to make an oxygen-production system. The objective of the model can, therefore, be stated as to find the output level, O, (which determines the size of the production machine), the compressor motor capacity, H, (which determines the size of the compressor), and the inventory storage capacity, V, (which determines the size of the storage unit) that will provide the minimum total cost of production per cycle, C. It is assumed that the cost equations of these three machines are known to be functions of O, H, and V respectively and the parameters in the equations are assumed to be unchanged for the future, and that the demand pattern

---

[9]The inventory storage unit is regarded as one "machine" in the context of our model. Indeed, an inventory storage unit can always be regarded as a "machine" that changes the delivery time of the product.

is cyclic. Thus, the problem can now be stated formally as:

Minimize $C = C_1 + C_2 + C_3$,

where C_1 (Production Machine Operating Cost) = $f_1(O;D,\alpha,\beta,\gamma)$,

C_2 (Inventory Processing Cost) = $f_2(H;D,\alpha,\beta,\gamma)$,

C_3 (Inventory Storage Cost) = $f_3(V;D,\alpha,\beta,\gamma)$,

D is a given cyclic demand pattern, and

α,β,γ are vectors of parameters in the cost equations

of the three machines, such as cost of the inputs,

units of the inputs used, etc.

Obviously, the optimal levels of the capacity variables, O, H, and V, depend not only on the demand pattern within the cycle but also on the parameters in the cost equation. Without the loss of generality, we will assume that the demand is of a cyclic two-step nature with a constant demand of D_0 for the period O to t_1, and a constant demand of D_1 for the period t_1 to t_2.[10] The minimum constant level of output for such a demand pattern is, of course, the average demand of $D_{av.}$. The demand function with $D_{av.}$ superimposed can thus be portrayed as in Figure 8-1.

The maximum inventory capacity required, I_m, must be:

$$I_m = (D_1 - D_{av.})(t_2 - t_1) = (D_{av.} - D_0)t_1 \qquad (8-1)$$

---

[10]Cyclic stepwise demand functions are common to the problem this paper analyzes, hence, this assumption is used. The technique presented is, of course, applicable to all kinds of cyclic demand functions.

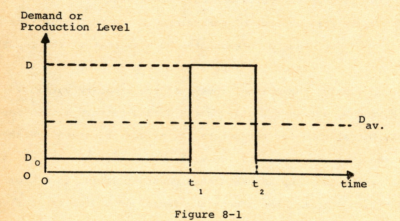

Figure 8-1

Stepwise Demand Function

while the minimum inventory capacity, I_{min}, is zero.

The Capacity Variables

Following the general model, the capacity variables can be expressed as functions of the decision variables and the other pre-determined parameters. The decision variables in the specific model consist of the oxygen facility production rate, P, and the maximum pressure of the storage capacity, p. The first capacity variable, O, is the output capacity of the oxygen production plant during the cycle:

$$O = g_1 (P;t_2) = Pt_2 \qquad (8-2)$$

where P is the oxygen production rate per time unit and t_2 is the length of the demand cycle in time units, a parameter.

The second capacity variable is the required volume of the storage facility, V; derived from the well-known gas law[11];

$$V = g_2 (p,P;G,R,T)$$
$$= (D_1 - P)(t_2 - t_1) GRT/p \qquad (8-3)$$

where the decision variable p is the maximum pressure of the compressor and the storage facility, P is as defined above, G is the gas compressibility factor, R is the gas constant, and T is the gas temperature in degrees Rankine.

---

[11] O.W. Eshbach, editor, <u>Handbook of Engineering Funda-mentals</u>, second edition, New York: John Wiley and Sons, Inc., 1958, pp. 8.15-8.19.

The third capacity variable, the horsepower required to operate the compressor is determined by its maximum instantaneous power. This power is required when the inventory approaches inventory capacity, I_m, and when demand is at some minimum level, D_o. Under these conditions, the maximum power is determined by the maximum pressure ratio p/p_o, where p_o is the production pressure. The driver horsepower required to, figuratively speaking, "put the last pound of gas into storage," is:

$$H = g_3(P,p;D_m,t_1,t_2,k_1k_2R,T,p_o)$$
$$= (D_1-P)(t_2-t_1)RT\ln(p/p_o)\Big/(t_1k_1k_2) \qquad (8-4)$$

where H is a function of both decision variables P and p. The parameters or known constants not already discussed are k_2, the overall isothermal compressor efficiency, and k_1, which transforms foot-pounds into horsepower-hours.

The constraints on the decision variables, P and p, are:[12]

$$P \geq D_{av.}$$

where $D_{av.}$ is the average demand rate, and

$$p \geq p_o$$

---

[12] Some readers may be wondering why the constraint, $P_t+I_{t-1} \geq D_t$, is not used. The reason is that, given a two-step cyclic demand pattern, $P \geq D_{av.}$ implies that this constraint is also met.

where p_O is the production pressure of the oxygen which
equals the minimum delivery pressure of the oxygen.

The Cost Functions

The cost functions are now presented. The oxygen production cost, C_1, including current outlays and depreciation
is:

$$C_1 = a_1 + a_2 O/t_2$$

where a_1 and a_2 are empirically-determined and known parameters. This function results from work by Katell and Faber[13]
which revealed that oxygen gas production above a certain
minimum production rate is linearly related to the plant production rate.

There are two kinds of costs associated with compressing
the oxygen. They are: the operating cost of the compressor
(excluding power) C_{21}, and the power cost C_{22}. For operating
cost, Chilton[14] indicated that the compressor installed cost
is an exponential function of driver horsepower, H. Assuming
that current outlays and depreciation cost is a fraction of
the installed cost, the operating cost of the compressor (excluding power) C_{21}, is therefore:

$$C_{21} = b_1 b_2 H^{b_3}$$

[13]S. Katell and J.H. Faber, "What Does Tonnage Oxygen
Cost?" Chemical Engineering, June 1959, pp. 107-110.

[14]C.H. Chilton, editor, Cost Engineering in the Process
Industries, New York: McGraw-Hill, 1960.

where b_1 is the fraction of the installed cost which converts the installed cost to operating cost per cycle, b_2 and b_3 are known parameters. The cost of the power consumed by the compressor in compressing the gas from production pressure p_o to the desired storage pressure p, C_{22}, is:

$$C_{22} = b_4 t_1 H$$

where b_4 is the cost of the power used per cycle.

Summing C_{21} and C_{22} yields the total inventory processing cost per cycle:

$$C_2 = C_{21} + C_{22} = b_1 b_2 H^{b_3} + b_4 t_1 H \qquad (8-6)$$

As to the inventory storage cost, Vilbrandt and Dryden[15] find that the installed cost of pressure vessels can be represented by an exponential function of the vessel volume, V. Assuming again that current outlays and depreciation cost can be expressed as a fraction of installed cost, the inventory storage cost, C_3, becomes:

$$C_3 = c_1 c_2 V^{c_3} \qquad (8-7)$$

where c_1 is the fraction of the installed cost which converts the installation cost to operating cost per cycle; c_2 and c_3 are parameters.

Summing up the cost equations (8-5), (8-6) and (8-7) and expressing the capacity variables O, V, and H in terms of the decision variables P and p according to equations (8-2), (8-3) and (8-4), yield the total cost equation to be minimized.

---

[15] F.C. Vilbrandt and C.E. Dryden, <u>Chemical Engineering Plant Design</u>, 4th ed., New York: McGraw-Hill, 1959.

$$\text{Min. } C = a_1 + a_2 P + b_1 b_2 \left[(D_1 - P)(t_2 - t_1) RT \ln(p/p_0) \Big/ (t_1 k_1 k_2) \right]^{b_3}$$
$$+ b_4 (D_1 - P)(t_2 - t_1) RT \ln(p/p_0) / (k_1 k_2) + c_1 c_2 \left[(D_1 - P)(t_2 - t_1) GRT/p \right]^{c_3}$$

Subject to
$$P \geq D_{av.}$$
$$p \geq p_0$$

METHODS OF SOLUTION

Since the total cost equation is a very complex non-linear function of the decision variables P and p, classical calculus techniques are inadequate. We will, therefore, use a reasonable set of numerical values of the parameters (see the Appendix) to obtain a numerical solution through a gradient method.

The Gradient Method

The gradient method starts with an arbitrary initial point (P_1, p_1). The values of (P_1, p_1) are then inserted into the partial derivatives, which will be denoted $f_P(1)$ and $f_p(1)$, and a new point (P_2, p_2) is determined by the formulas:

$$P_2 = P_1 + h f_P(1) \Big/ \left[f_P^2(1) + f_p^2(1) \right]^{\frac{1}{2}}$$
$$p_2 = p_1 + h f_p(1) \Big/ \left[f_P^2(1) + f_p^2(1) \right]^{\frac{1}{2}}$$

To compute (P_3, p_3) the values of (P_2, p_2) are used and this recursion is continued until a stationary point is reached. The parameter h is a numerical value. This value may be held constant or it can be varied according to the magnitude of $f_p(i)$ and $f_p(i)$. The selection of h is fairly critical. If h is chosen too large, a stationary point may never be reached, while if it is chosen too small, too many steps will be required to reach a stationary point.

The gradient procedure is fast and efficient, especially when a satisfactory value for h has been determined. However, one serious disadvantage of the method is that only stationary points are found, which may or may not include the global minimum. Hence the procedure must be repeated with several starting points, and each stationary point obtained is then evaluated in the total cost function.

The Solution

The gradient procedure was applied to the problem based on the parameters in the Appendix. The solution is presented in Table 1, which shows that a border solution for P, its lower limit, was obtained. As can also be observed from Table 8-1, cost is very insensitive to the variable p. Pressure can be varied considerably without affecting the cost to any significant extent. However, one would expect that if the cost of power were changed the optimal value of p would change also. Since p determines the compressor capacity and the storage capacity configuration (the volume) it is important that information regarding the effects on p of varying the power cost is available. With this in mind, several gradient runs were made on the computer with varying power costs. The results of these runs are shown in Table 8-2, and indicate that the optimal p is significantly changed if the power cost decreases or increases. Hence, knowledge of future power rates is an important input to the decision problem herein considered.

Table 8-1
Solution Set For Example

Production Rate	Max. Inv. Pressure	Cost
35.00 tons/hour	520 psia	$265.51
28.00	500	250.00
24.00	497	208.11
20.00	495	187.04
17.50	490	173.834
17.50	483	173.831
17.50	475	173.829
17.50*	461*	173.828*

* Minimum cost solution based on power cost of $.006/(HP-HR).

Table 8-2

Effect of Power Cost on Maximum Inventory
Pressure

Power Cost	Production Rate	Max. Inv. Pressure	Minimum Cost
$.0015/(HP-HR)	17.5 tons/hour	783 psia	$172.21
.003	17.5	587	172.85
.006	17.5	461	173.83
.009	17.5	351	174.52
.012	17.5	278	174.96
.018	17.5	200*	175.07
.024	17.5	200*	175.07

* 200 psia is the minimum inventory pressure allowed.

EXERCISES

1. Suppose the stepwise demand function in Figure 8-1 is changed to a repetitive linear demand function of the form,

$$D(t) = D_0 + (D_1 - D_0)t / t_2$$

where t_2 is the length of the cycle, D_0 is still minimum instantaneous demand and D_0 is still maximum instantaneous demand. What would now be the maximum required inventory capacity? How would the capacity variables O, H and V change? What is the form of the total cost function?

2. Repeat exercise 1 with a positive exponential function of the form,

$$D(t) = \exp(\alpha t)$$

where $D_1 = \exp(\alpha t_2)$, $D_0 = \exp(0)$ and t_2 is length of cycle.

3. What other solution method could be used to solve for optimal pressure, p, and production rate, P, in the oxygen production process problem discussed in this chapter?

4. A common problem in pipeline construction planning is the selection of the optimal combination of pipe diameter and booster stations to maintain the required flow in a pipeline. Smaller diameter pipe needs more booster capacity. Determine a capacity variable function for the pipe and the booster station in terms of decision variable vector X_i and a parameter vector Z_i.

5. Using the general model outlined in this chapter specify the cost functions for each of the capacity variables specified in the previous problem. Complete the optimization problem by specifying constraints. Suggest solution method for your model.

6. Suppose the pipeline construction problem presented above has the added alternative of providing a special surface finish to the inside of the pipe which will reduce friction and increase capacity. The amount of friction reduction is, however, a function of pressure in the pipe and the diameter of the pipe. How would you include this added variation in your optimization model?

APPENDIX

List of Parameters and Constants

symbol		value	dimension	relates to
a_1	=	61.8 ------------dollars/hour		C_1
a_2	=	5.72 -----------dollars/ton		
c_1	=	.000025-------dollars/hour		C_3
c_2	=	376--------------$(cubic\ feet)^{-b_3}$		
c_3	=	0.75-----------dimensionless		
G	=	0.98-----------dimensionless		V
R	=	670.6-----------cubic feet-psia/(or tons)		
T	=	530----------------$^\circ R$		
b_1	=	.000025-------dollars/hour		C_{21}
b_2	=	700-------------- $(HP)^{-c_3}$		
b_3	=	.85-----------dimensionless		
D_0	=	2.5-------------tons/hour		C_{22}
k_2	=	0.70----------dimensionless		
k_1	=	14005.8-----------cubic feet-psia/(HP-HR)		
p_0	=	200---------------psia		
b_4	=	.006----------dollars/(HP-HR)		
D_1	=	40-------------tons/hour		
t_1	=	0.6-----------hours		
t_2	=	1.0-----------hours		

CHAPTER 9.--CAPACITY PLANNING-BASED ON ENDOGENOUS RELATIONSHIPS[1]

This chapter will concentrate on a technique for determining the optimal machine characteristics for a machine which fabricates two-dimensional products. The machine is a high-speed, high-cost, continuous processing facility. The end products of this continous operation, called intermediate inventory, are stored and then used as input for a low-speed, low-cost, job-shop operation which finishes them to customers' dimensional specifications.

The method used to resolve the above problem is similar to but different from methods used in the previous two chapters. The difference lies in the methods used to derive the relationship between the decision variables. In the previous two chapters underlying but exogenous engineering relationships were used to derive the production and capacity functions. The problem in this chapter has no basic underlying engineering relationships that can be used to derive the necessary relationship functions. Therefore, simulation methods were used to generate sample data which were then used to develop analytical cost functions. These cost functions consist of mathematical relationships between the endogenous decision variables, the machine characteristics.

---

[1] This chapter is based on C. Carl Pegels: "A Technique for Determining Optimal Machine Characteristics," _The International Journal of Production Research_, Vol. 6, No. 1, 1967, pp. 47-56.

PROCESS DESCRIPTION

The optimum machine characteristics determined by the model can be called width capacity and length capacity. The intermediate product is created as a continuous strip which can be thought of as having a defined width and an infinite length. The machine is capable of making this strip in any of a number of widths, but the widest strip cannot exceed the width capacity of the machine. The question is: what should this width capacity be? A machine with large width capacity is expensive to acquire but may provide low unit operating cost. A machine with narrow width capacity is less expensive to acquire but will have a higher unit operating cost and a higher waste cost.

Before the strip leaves the machine, it is cut into finite lengths. As with widths, the machine is capable of making any of a number of length cut-offs, and a similar cost issue arises. Very long lengths of intermediate product will cause little waste, but are expensive to handle because special material handling equipment must be acquired. Shorter lengths will be less expensive to handle but will incur a higher waste cost when cut into lengths specified by the customers' orders.

METHOD OUTLINE

Three different techniques are used to solve the problem. These are simulation, multiple regression and optimization.

Simulation

   Simulation provided a model of the system, which could
be used to generate observations on output data as functions
of the input data. The input data variables that were varied
were machine width capacity and intermediate product length.
The output data observed consisted of waste incurred and
number of machine hours required to produce a unit quantity
of product.

Multiple Regression

   The data generated by the simulation model have been
summarized in a set of functional relations by using multiple
regression. These functional relations have waste incurred
and machine hours as dependent variables, and machine width
capacity and intermediate product length as independent variables.

   It is assumed that cost data are available from which
waste cost per unit quantity and machine cost per time period
can be estimated. A cost function will be derived by combining the cost data and cost functions with the functional relations obtained by regression. Optimization techniques
will then be applied to solve for the minimum cost values
of machine width capacity and intermediate product length.

Simulation and Analytic Techniques

The reader at this point may wonder why not use simulation solely, instead of a combination of simulation and analytic techniques. Or alternatively, he may ask why not use analytic techniques to solve the complete problem.

To use only analytic techniques would require simplification of the system being studied to the point where the analytic solution would be practically useless. For instance, the system being simulated has three stages and twenty-five product lines, and items in one customer order ordinarily have different physical shapes from items in other orders in the same product line. Operating cost and waste cost are determined by the interactions of the orders, the first stage, by the way the orders are scheduled for production, the second stage, and by the availability of raw material and other production considerations, the last stage. Although the system just described may theoretically be formulated in an algebraic equation, to do so would require numerous simplifying assumptions, and the final results could be seriously affected by these assumptions.

Simulation abandons the idea of expressing the problem by a general algebraic statement, but in exchange one gains the advantage of a more realistic model of the system being portrayed.

Using only simulation to solve the complete problem
is feasible but not very practical. To do so requires that
the simulation model be provided with cost data and cost
functions so it can generate total cost output data. Any
feasible set of input data could then be simulated and a
total cost figure would be obtained for each set of input
data. Since we are   interested in the effect on total
cost of the input variables machine width capacity and intermediate product length, only those input variables would
have to be varied. The problem that arises is how to adjust
the input variables machine width capacity and intermediate
product length so that total cost moves in the direction of
its minimum point. Since no algebraic expression of the cost
function is available, the algebraic gradient method cannot
be applied. Cost slopes cannot be estimated from simulation
runs because of the variability of each individual run. Hence
the only method remaining is random search in the vicinity
of the expected minimum cost point.

In addition to the disadvantages of the simulation approach discussed above there is also the disadvantage of cost.
Computer simulation runs are relatively expensive in comparison with multiple regression and numerical optimization rou-
tines.

Figure 9-1 presents a graphical picture of the results
that will be obtained by the proposed method and those that
would be obtained by a simulation only approach. The curve

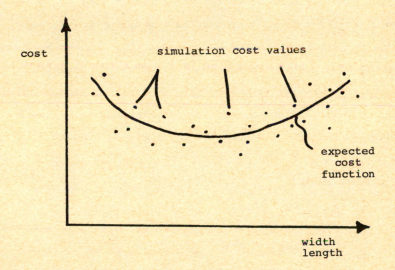

Figure 9-1
Total Cost Function

represents the cost function derived with the proposed method, and the points represent typical cost values obtained by simulation only.

SIMULATION MODEL

The simulation model represents a system consisting of a production operation, a production scheduling operation, and the market operation which purchases the finished product. The production process is two-stage. The first operation converts raw material into a continuous strip of product which is cut into large flat sheets, called intermediate product. The second operation finishes the intermediate product.

Production Planning

The planning of the two production operations is accomplished with one schedule. For each order, or set of two orders, the schedule prescribes the intermediate product length of the first operation. Intermediate product width could also be determined, but this model did not do so. The schedule then prescribes how the intermediate product should be finished on the second operation to customers' dimensional specifications.

Market Operation

The market operation generates orders and their respective specifications by random sampling from empirically-based distributions. The number of orders, the number of items in an order, the width of each item, and the length of each item are determined for each product line. Product lines are scheduled in a specified sequence that is based on the empirical operation.

Dependent Variables

The simulation model computes the amount of waste incurred, the percentage downtime for set-up, and the machine operating hours of the first production operation.

Waste is caused by various sources. The strip of product produced by the first operation has rough edges which must be trimmed smooth, and quantities of low quality product occur when rolls of raw material are changed or when the product line is changed. Waste is also produced when customers' orders are cut from the intermediate product. A typical intermediate product blank is shown in Figure 9-2. One, two, or three customer orders are scheduled to be cut from this blank. The shaded portions of each blank indicate waste. It can be observed that large blanks produce a lower percentage waste than small blanks. If only one customer order is

210

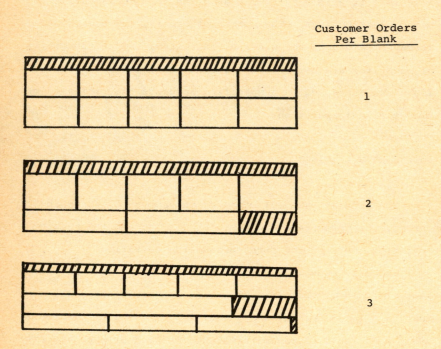

Customer Orders
Per Blank

1

2

3

Figure 9-2

Intermediate Product Blanks

cut from the blank, no end waste is incurred but side waste
is relatively high.

Downtime for set-up is caused by product line changes,
raw material changes, and miscellaneous adjustments. In
most cases a considerable portion or all of the set-up work
is done while the machinery is running, because the machinery
is set-up to run an order while the previous order is being
fabricated. If short orders are encountered there is insufficient time to do all the set-up required for the next
operation, and the remaining portion which is done while the
machine is shut down, is charged to set-up downtime. Raw
material changes, which are caused by a change of product
line are similarly charged for set-up downtime if any portion of the change over set-up cannot be done while the machinery is in operation. Miscellaneous set-up for special
orders is ordinarily done while the machinery is shut down
and hence is charged to set-up downtime.

Machine hours include both operating time and set-up
downtime. Machine operating hours are determined by the
product line being run. There are three weight classes of
product lines. Each weight class is fabricated at a different machine speed, which is fast for light weight product lines, normal for medium weight product lines and slow
for heavy weight product lines. Set-up for the next order
is assumed constant in the model, hence an order for a light
weight product line will incur more set-up downtime than an

order for a heavy weight product line of equivalent size.

COST FUNCTIONS

The cost function[2] to be minimized is comprised of four cost components. These are waste cost, machine operating cost, machine capital cost, and material handling capital cost.

Waste Cost

Expected waste cost is determined by the expression:

$$E(WC) = k_2 f_1(W,L),$$

where k_2 is raw material cost per unit area; W and L are the machine width capacity and length of intermediate product respectively; and $f_1(W,L)$ is the functional relation between waste, machine width and intermediate product length.

Operating Cost

Expected machine operating cost is determined by the expression:

$$E(OC) = g_1(W) \cdot f_2(W),$$

where $g_1(W)$ is the machine operating cost per hour. It is

---

[2] For a discussion of cost functions and their estimation see Davidson, R.K., V.L. Smith, and J. W. Wiley, Economics: An Analytical Approach. Revised Edition, Homewood, Illinois: Richard D. Irwin, Inc., 1962.

assumed that it can be estimated for machinery manufacturer's data. Its functional form is assumed to be:

$$g_1(W) = b_{10} + b_{11}W.$$

Both coefficients are positive. A linear function is assumed because the rate of cost increase is expected to be constant between different machine capacity widths. The function $f_2(W)$ is the functional relation between operating hours per unit quantity of product and machine width capacity.

Capital Cost

The machine capital cost is expressed as cost per unit quantity of product. This is a deviation from the usual custom of expressing capital cost as cost per unit time. However, if annual production figures are known, one function can be derived from the other. The machine capital cost per unit of product is a function of width only; it is expressed as:

$$E(CC) = b_{20} + b_{21}W + b_{22}W^2. \qquad (9-1)$$

The material handling equipment capital cost depends only on the length of the intermediate product. Different widths are not expected to have significant effects. It is expressed as:

$$E(HC) = b_{30} + b_{31}L + b_{32}L^2. \qquad (9-2)$$

The coefficients of the cost functions (9-1) and (9-2) are assumed to be positive for the intercepts, negative for the linear term and strongly positive for the quadratic term. These coefficents are assumed because capital cost is expected to increase at an increasing rate as width capacity and length of intermediate product move above certain levels.

Prediction Equations

The prediction equations $f_1(W,L)$ and $f_2(W)$ have been estimated from data generated by simulation runs. Waste incurred and operating hours per unit quantity of product are expected to decrease, as width and length increase, at a declining rate as pictured in Figure 9-3.

The functions that fit this curve are:

$$f_1(W,L) = K_0 + K_2(W + K_3 L - K_1)^2 \qquad (9-3)$$

$$f_2(W) = K_0 + K_2(W - K_1)^2 \qquad (9-4)$$

K_0, K_1, K_2 and K_3 are constants and K_1 is the upper bound of the region analyzed.

Expanding these equations[3] results in the following postulated forms which have been used to estimate the prediction equations for waste and operating hours:

---

[3] Equations (9-3) and (9-4) become $f_1(W,L) = (K_0 + K_2 K_1^2) - 2K_1 K_2 W - 2K_1 K_2 K_3 L + K_2 W^2 + K_2 K_3^2 L^2 + 2K_2 K_3 WL$, and $f_2(W) = (K_0 + K_2 K_1^2) - 2K_1 K_2 W + K_2 W^2$.

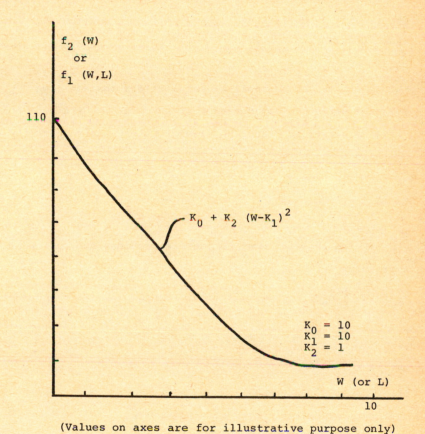

(Values on axes are for illustrative purpose only)

Figure 9-3

Expected Functional Relation

$$f_1(W,L) = a_{10} - a_{11}W - a_{12}L + a_{13}W^2 + a_{14}L^2 + a_{15}WL,$$

$$f_2(W) = a_{20} - a_{21}W + a_{22}W^2.$$

Analysis of variance tests indicated that the coefficients a_{12} and a_{14} were not significant, and were therefore deleted. The remaining coefficients were significant and the signs of these coefficients agreed with the coefficients derived from equations (9-3) and (9-4) with the exception of a_{15} which turned out to have a small negative effect instead of a positive effect. This one exception appears to indicate that waste decreases at a somewhat faster rate than indicated by equation (9-3).

Total Cost Function

Adding the four cost components provides the following expression for combined cost:

$$E(TC) = c_0 + c_1W + c_2L + c_3W^2 + c_4L^2 + c_5WL + c_6W^3. \tag{9-5}$$

subject to: $a_1 \leq W \leq b_1$

$\qquad\qquad a_2 \leq L \leq b_2$.

The upper and lower bounds on the dimensions are determined by technological and physical constraints.

Optimization

The usual technique for finding the optimal value

of W, W_0, and the optimal value of L, L_0, is to set the partial derivatives, $f_W(W,L)$ and $f_L(W,L)$ of the cost function (9-5) equal to zero. The gradient method[4] can then be applied to the two resulting equations to solve for W_0 and L_0. The disadvantage of the gradient method is that each series of calculations, from a starting point, only results in a stationary point, which may or may not be the global optimum. Therefore, a more direct method, which will be called constrained global optimization, has been used.

Constrained global optimization is an efficient and rapid way to find the global optimum of this specific example problem. The optimization technique is applied by substituting feasible values for all but one of the variables in the equations obtained by setting the partial derivatives equal to zero. The resultant equations which must be polynomials or transformable into polynomials can be solved by numerical methods. Numerical methods become quite costly for higher-degree polynomials and hence constrained global optimization is more suitable for problems which have at least one variable with a low degree. In addition, for more than two variables the number of computations required becomes phenomenal. For instance, if n calculations are required for a two-variable problem, then n^{m-1} calculations are required for an m-variable problem. Although constrained global optimization appears to be very efficient on the example problem, it will not be on more complicated problems.

---

[4]For a discussion of the gradient method see Carr, C.R. and C.W. Howe, Quantitative Decision Procedures in Management and Economics, New York: McGraw-Hill, 1964.

However, the limited class of problems, which can be solved efficiently by constrained global optimization, probably includes a considerable portion of the practical problems encountered in industry.

A two-variable problem, such as the example problem presents three possible polynomials; any of the three polynomials can be used to solve the problem. In general there usually will be one method that is preferred depending on the amount of computation required.

The three cases are:

Case 1: $f_W(W,L) = 0$ (9-6)

Case 2: $f_L(W,L) = 0$ (9-7)

Case 3: $f_W(W,L) - f_L(W,L) = 0$ (9-8)

If a feasible value for one of the variables is substituted in the above equations, then each equation can be solved for the value or values of the other variable. If the sets of values obtained for both variables are feasible, then these sets of points may belong to the set of stationary points and can be evaluated in the cost function (9-5). Suppose W is the variable that is substituted , then one way to attack the problem is to divide the range of W, $a_1 < W < b_1$ into small equal increments, so that $a_1 < \bar{W}_1 < \bar{W}_2 < ... < b_1$. Each $\bar{W}_i$ is inserted into equations (9-6),(9-7) or (9-8), and the equation is solved for L. All feasible points and border points are evaluated in the cost function (9-5) and the minimum cost point is found. The example problem was solved by this method. Equal increments of W were used although variable increments especially near the optimum may in some cases be more desirable.

For problems such as the example problem constrained global optimization takes about twice as long on a computer as one iteration of the gradient method. However, each iteration of the gradient method only produces a stationary point, which may or may not be the global optimum.

AN APPLICATION

The above methods were applied to the problem for which the simulation model was designed. Twenty simulation runs were made with different sets of width capacity and intermediate product length to generate sufficient data which were used to obtain functional relations between input and output data by multiple regression techniques. Cost functions were estimated and combined with the functional relations. The result was a combined cost function[5] in terms of machine width capacity and intermediate product length.

Partial derivatives were taken with respect to width and length, and the constrained global optimization method applied. The variable W was taken in increments of 1.2 units from the lower bound of 78 units to the upper bound

---

[5] Its coefficients are: $c_0=1189.31$, $c_1=34.919$, $c_2=-3.715$, $c_3=1.645$, $c_4=.01$, $c_5=-.001434$, and $c_6=.0183$.

of 150 units. The variable L was constrained between 150 and 480 units. The optimum machine capacity is presented in Table 9-1.

The results indicate that the optimum width is considerably wider than the minimum width of the machine. The reduction of waste cost and operating cost per unit quantity by operating a machine with a wider width was cancelled by an increase in operating cost per unit time. Hence, if machinery could be provided with lower operating cost per unit time the maximum width would become attractive. A result of this sort provides hints where cost reduction efforts should be applied. The high cost of handling long intermediate product blanks caused the optimum length to be considerably below the upper bound. Again management is informed by these results that cost reduction efforts should be applied to the handling of intermediate product.

CONCLUSIONS

A model has been presented for determining the optimal capacity of a machine which fabricates two-dimensional products on a high-speed, high cost continuous processing facility, and finishes these products on a slower-speed, lower-cost processing facility. The model was developed for a specific application in the paper industry, but problems of

221

Table 9-1

Optimum Machine Capacity

	Case 2: $f_L=0$			Case 3: $f_W=f_L$	
cost	length	width	length	cost	
$690.10	186.22	78	infeasible	--	
684.86	186.25	84	"	--	
680.64	186.29	90	"	--	
677.44	186.32	96	"	--	
675.29	186.36	102	"	--	
674.19	186.40	108	"	--	
674.05	186.41	110.4	173.74	$675.66	
674.05*	186.42*	111.6*	193.83*	674.60*	
674.08	186.42	112.8	213.98	681.68	
674.22	186.43	114	234.18	696.96	
675.22	186.47	120	335.93	898.61	
677.37	186.50	126	438.97	1314.75	
680.64	186.54	132	infeasible	--	
685.02	186.57	138	"	--	
690.55	186.61	144	"	--	
697.23	186.65	150	"	--	

*minimum cost solutions by each method

the same basic type are encountered in the manufacture
of products of steel, aluminum, and so forth.

Several techniques were used to solve the problem,
including simulation of hypothetical machine capacities,
multiple regression to determine functional relations and
a constrained global optimization technique. An example
problem was worked out to exhibit an application of the
model.

Other Applications

The methodology presented in this chapter may be used
to solve optimal machine capacity problems in other related
production situations. For instance, the optimum machine
capacities of a series of three continuous operations
could be obtained. Suppose the first operation's machine
capacity is determined by w and v. These parameters could
be width and length capacity or width and speed capacity
or other combinations. Assume the second operation requires two machines, because it is a slower process. The
capacities of the two machines are determined by w and x.
The third operation requires one machine and its capacity
parameters are w and y. Hence a cost model could be
built that would incorporate the four variables of the
three operation, four-machine production process. It is
assumed that functional relations between operating hours

and the variables can be obtained.

Applications to Discrete Operations

So far only continuous operations have been considered. However, the model could be applied to discrete operations as well. Suppose a plant receives
orders for a product which is made out of two-dimensional
raw material and fabricated to a three-dimensional finished product. Assume that product is demanded in various
sizes. The firm has to turn down orders that specify
sizes which exceed the maximum capacity of the machinery.
Let the first and second operations' capacities be determined by w and v, and the capacity of the third operation by w, and h. Hence, a three-variable, three-machine
model is required. In addition to operating, waste and
capital cost, the model would also consider the opportunity cost of lost sales for orders that cannot be processed on the machinery. Hence, the basic model discussed
in this paper can be generalized to have wide application
to many optimal machine capacity problems .

EXERCISES

1. Using the data in Tables 3-1 and 3-2 of Chapter 3, plot the percent downtime for set-up and the machine hours per million square feet as functions of change-over allowances and machine operating speeds. Notice the variability in the outputs of fairly lengthy simulation runs and re-estimate the values on the basis of a regression line.

2. Using the two regression lines calculated in the previous exercise and the data in Table 3-1, calculate the standard deviations around these regression lines. For a change-over allowance of 125 percent of validation run determine the 95 percent confidence interval estimates for percent downtime for set-up and for machine hours per million square feet.

3. Using the two regression lines calculated in exercise 1 and the data in Table 3-2 calculate the standard deviations around these regression lines. For a machine operating speed of 120 percent of validation run determine the 95 percent confidence interval estimates for percent downtime for set-up and for machine hours per million square feet.

4. Which other functional forms could have been hypothesized with reasonable justification for the waste function $f_1(W,L)$ and the machine hours function $f_2(W)$?

5. Apply the constrained global optimization technique discussed in this chapter to find the maximum and minimum of the function,
$$g(x,y) = a_0 + a_1 x + a_2 x^2 + a_3 x^3 + a_4 y + a_5 y^2 + a_6 y^3 + a_7 xy$$

for a specified two-dimensional plane of x and y.   Substitute reasonable numerical values for the coefficients.

6. The three-machine optimization problem discussed in this chapter has
the following controllable variables. For the first operation the machine
variables are width (w) and length (v); for the second operation two
machines are required with variables width (w) and speed (x); the third
operation requires one machine and its capacity variables are width (w)
and height (y). Develop a method whereby this optimization problem can
be resolved assuming that functional relationships between the variables
and cost parameters can be determined.

7. How would you apply the global optimization technique to find the
solution vector (x,y,w,v) to problem 6?

PART III. -- PLANT LAYOUT

PLANNING MODELS

Plant layout is one area in managerial planning where systems simulation, so extensively discussed in the previous chapters, has been widely used. However, the systems simulation used in plant layout has been a physical simulation, on a micro scale, of the equipment, machinery, aisles, work places and offices. As a matter of fact micro-scale simulation is so much a part of plant layout planning that few layouts are approved unless they have been evaluated by micro-scale simulation.

It is not the intention to discuss the pros and cons or the techniques of micro simulation in the next three chapters. What is planned, however, is a discussion of various quantitative techniques which will supplement and greatly improve the plant layouts now determined by the common classical approaches. The proposed techniques all have one common objective. That objective is to develop plant layouts which will incur a minimum of material handling cost. As one executive has pointed out, "The cost of moving material from one manufacturing area to another is a complete loss insofar as its effect upon the value of the product is concerned. For this reason it should be considered a manufacturing evil that must be completely eliminated or reduced to an absolute minimum"[1] (Cameron, 1952).

In Chapter 10 a brief review of plant layout will be given. The common methods of travel charting will be discussed and methods will be presented whereby improved layouts can be developed. Heuristic problem solving will be introduced.

---

[1] Cameron, D. C., "Travel Charts – A Tool for Analyzing Material Movement Problems," Modern Materials Handling, Volume 8, No. 1.

Chapter 11 presents an approach for determining a plant layout by a heuristic method which has been computerized. This method, also known as CRAFT, will be described and simple illustrations will be provided. Buffa, et al[2] have pointed out several applications of the CRAFT plant layout computer program, discussed in Chapter 11. In one case CRAFT applied to a large integrated movie studio indicated a fantastic cost reduction of nearly $240,000 per year in material-handling savings alone through the relocation of many of the major functional departments. Another application of CRAFT in the location study of a special laboratory testing service department resulted in a cost reduction of 14 percent by showing that the location should be centralized. Still another study of a maintenance department resulting in a material handling cost reduction of nearly 27 percent.

Chapter 12 presents a heuristic/optimizing procedure for plant layout which holds great promise because it continues to search for the best solution until the cost of computing exceeds the expected additional benefits to be derived from further search.

Other potential applications of the models discussed in the next three chapters are in non-manufacturing situations. For instance, optimal office layout will minimize the travel of personnel required for face-to-face communications. Similarly, optimal location of goods or parts in a warehouse will minimize the cost required to move these goods or pick up the parts when required. In a retail store or supermarket the location of the various departments

---

[2] Buffa, E. S., G. C. Armour, and T.E. Vollmann, "Allocating Facilities with CRAFT," Harvard Business Review, March-April 1964.

can have a significant effect on costs of transporting and replenishing inventory.
Finally, in a hospital, the location of the X-ray room, blood bank, laboratories,
store rooms, kitchen and operating rooms all affect the cost of operating the
hospital.

CHAPTER 10.--THE BASIC PLANT LAYOUT PROBLEM

The plant layout problem can be viewed as consisting of two parts. First, there is the problem of arranging the production and service departments in the plant. Secondly, there is the problem of arranging the machinery, equipment, aisles and work places within a department. Although these problems may interact to a certain extent in most applications the two problems can be analyzed, studied and resolved separately.

The main criterion that will be considered in this chapter and the subsequent two chapters is the development of methods to determine departmental layouts which will minimize cost of moving materials from one department to another. The same methods may, however, also be used to determine the best locations of machinery and equipment within a department.

OBJECTIVES

A plant layout is an arrangement of facilities and services in a plant. Its main objectives are to integrate the various production departments, and within the production department the production units into a logical, balanced and effective production operation to ensure that material flow between departments and between production units is smooth and efficient.

A more general statement of a plant layout objective
is the development of an arrangement of machines, men, materials and supporting services so that total cost of the
production system is minimized subject to the requirement
that the needs of the people associated with the production
system are satisfied.

Numerous other objectives may be identified. However,
the topic of plant layout in this chapter will be restricted
to the analysis of the main objectives stated above. It
is realized, however, that restriction of the analysis as
described above invites the accusation that suboptimization
is taking place.

PROCESS AND PRODUCT LAYOUTS

There are two basic production systems or layouts.
One is the process or functional layout. The other is the
product production system or product layout. In addition
to the two basic layouts there is also the combination of
process and product layouts.

Process Layout

The best example of a process layout is the machine
shop which operates on a job shop basis. The layout of a
machine shop includes separate areas for the various
machinery operations such as milling, drilling, grinding,
turning, heat treating and finishing. Another example of

a process layout is a woodworking shop which has areas set apart for shaving, planing, drilling, turning, gluing, and assembling. In both examples the different production processes provide the basic divisions of the plant layout.

Process layouts are commonly found in smaller operations which usually produce products on a custom-made basis and in limited quantities. The equipment and machinery found in a process production system are usually of a general-purpose type and therefore cheaper than the specialized equipment found in product layouts. Machinery breakdowns are of little concern in the process layout since the particular operation can usually be switched to another machine of the same or similar type without undue cost or trouble.

Product Layout

Mass-production techniques and automation have produced a wide variety of product production systems or product layouts. There are essentially two kinds of product layouts, the continuous, sometimes automated, production line on the one hand and the somewhat more old-fashioned line of independently-functioning machines on the other hand. The automated production line is gradually replacing the older independently operating machines. However, special purpose machines as found on automated

lines are extremely expensive and are therefore only
economical if very high volume is required.

Another kind of product layout is the assembly line.
It may be highly sophisticated or rather simple. Its products may all be exactly alike or it may produce many
variations of basically the same product. For instance,
automobiles coming off an assembly line are all basically
the same but the probability of two identical automobiles
coming off the line in an hour may be very small.

Disadvantages of product layouts, especially of
automated lines are problems caused by breakdowns of individual work stations on the line. If one work station
breaks down the whole line could be down shortly there-
after. To counteract this problem interstage storage is
frequently provided to reduce line stoppages. Hence, the
benefits of automated lines must be weighed against the
cost of individual work station breakdowns.

Combinations of Process and Product Layouts

Although process and product layouts are the way to
identify processes there are many in-between production
processes which are really combinations of process and
product layouts.

The advantage of the combination type of layout is
that one gains the benefit of product layouts in that the
product moves from work station to work station and also

the benefit of process layouts in building up adequate in-
process inventories which reduce the cost of individual
work station breakdowns.

A typical example of a combination type of layout
is found in a production department which produces a gear
box assembly. The machines for machining the gear box,
the gears and shafts are all independently operated. For
each operation more than one machine is commonly available
to do the work and the in-process inventory moves through
the department on independent means. The above description
of the gear box production process is neither the most
efficient nor the most up-to-date for high volume production. If high-volume production was required then an automated product production process would be desirable. However, for limited volume the above-described layout offers
many advantages.

MATERIAL HANDLING CRITERION

The main objective of this section is to determine
an efficient layout on the basis of minimum-cost operation.
The costs that will be considered include material handling
and departmental location costs. The last item assumes
that a cost item can be imputed to department location and
will be discussed in more detail in chapter 12.

By limiting the discussion to the above constraints,
a model will result which is analytically tractable, and

solvable. The limiting of the study by no means implies
that many of the other considerations in a plant layout are
considered unimportant. On the contrary, they are viewed
as being as important as the ones that are dealt with in
the models. Unfortunately, it is impossible to include all
objectives in one model and come up with a model that is
tractable and solvable. Looking at part of the model will
provide a solution which is based on many assumptions.
These assumptions may, however, be varied and the model
may be experimented with to try out various solutions
based on differing assumptions.

Best Relative Location of Department

The major problem to be analyzed in this and the
next two chapters is the determination of the most efficient relative location of the various production departments. Only for very simple layouts is it possible to
obtain the best arrangement by inspection or enumeration.

For more complex layouts the combinatorial problem
grows to enormous proportions. For instance for only six
equal size rectangular areas, as shown in Figure 10-1,
there are 6! or 720 layouts possible. However, only 45
of the total possible layouts are different if material
handling cost is considered as the major criterion in the
objective function.

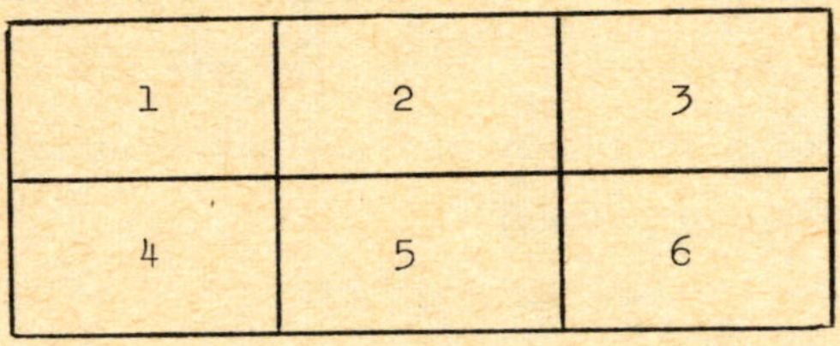

1	2	3
4	5	6

Figure 10-1

Simplified Layout of Six Equal Size Departments

Referring to the sample problem, a layout is required that places the departments in such relative locations to each other, so that material handling cost for all material handling in the plant is minimized. For instance, if total material handling between departments 1 and 3 was double the material handling between departments 1 and 2, then a logical rearrangement would be to interchange departments 2 and 3. However, this rearrangement would only be an improvement if it did not increase material handling between other departments by more than the decrease resulting from the above interchange. Hence, any potential interchange should be analyzed to determine its total effect on material handling.

Loads and Distance Matrices

To obtain the quantitative measure to be minimized two sets of data are required. These are the number of loads that are to be moved between departments during a given time period, and the distances between departments in the plant. These two sets of data can be neatly summarized in the two matrices which will be called A and D and are shown in Tables 10-1 and 10-2.

Matrix A contains the loads data, and indicates the number of loads which have to be moved from the department in each row to the department in each column. Similarly, matrix D indicates the distance that must be traveled between the departments in the rows and the departments in the columns. However, note that matrix D only provides the distances for a given layout, specifically for the example layout shown in Figure 10-1.

Table 10-1

Load Summary - Matrix A

Depts.	1	2	3	4	5	6
1	-	30	-	15	-	40
2	-	-	45	-	50	-
3	10	20	-	50	-	5
4	-	-	-	-	25	-
5	25	-	25	10	-	60
6	-	10	-	-	40	-

Table 10-2

Distance Summary - Matrix D

Depts.	1	2	3	4	5	6
1	0	8	16	6	10	17
2	8	0	8	10	6	10
3	16	8	0	17	10	6
4	6	10	17	0	8	16
5	10	6	10	8	0	8
6	17	10	6	16	8	0

Note that the load summary in Table 10-1 can be summarized in a triangular matrix as shown in Table 10-3. This summarization assumes that the cost of material handling in either direction between departments is the same. The distance summary can also be reduced by just deleting the lower triangular half of the matrix and using just the upper triangular portion.

Table 10-3

Load Summary in Triangular Form

Depts.	1	2	3	4	5	6
1	0	30	10	15	25	40
2		0	65	0	50	10
3			0	50	25	5
4				0	35	0
5					0	100
6						0

Using the two matrices represented by Tables 10-2 and 10-3 the value of the objective function can now be calculated by taking the number of loads of each A matrix element and multiplying it by the distance shown by the same D matrix element. Alternatively, if each element in matrix A is identified as a_{ij}, and each element in matrix D is identified as d_{ij}; then the value of the objective function, C, can be stated mathematically as

$$C = \sum_{j=1}^{n-1} \sum_{k=j+1}^{n} a_{jk} \, d_{jk} \qquad\qquad (10\text{-}1)$$

Improvements in Layouts

The solution to the above problem can be quickly calculated and is found to be C = 4550. To improve the solution the value for the other 44 possible layouts could be found and then the minimum value for the optimum solution could be selected. However, with a simple, or even a slightly more complicated layout a graphical layout as shown in Figure 10-2 can be used. Each department center, indicated by a small circle is located relative to the location of the other departments. The values shown on the connecting lines are the loads to be moved between departments for a given time period.

The question that arises is how can the solution be improved by rearrangement? A cursory inspection indicates that the initial layout is not a bad layout. However, it appears that an improvement may be obtained by moving departments 3 and 4 closer together. This may be accomplished in various ways, but it appears that the logical move is to interchange departments 3 and 5. Interchanging departments 3 and 5 generates the new distance matrix D shown in Table 10-4.

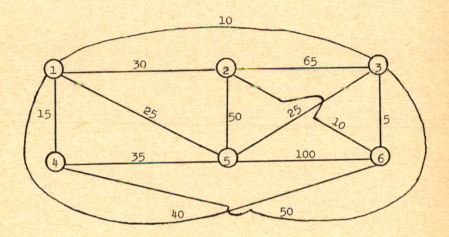

Figure 10-2

Schematic Diagram of Loads

Table 10-4

Revised Distance Summary

Depts.	1	2	3	4	5	6
1		8	10	6	16	17
2			6	10	8	10
3				8	10	8
4					17	16
5						6
6						

Applying formula (10-1) to the revised layout provides a value for the objective function of C = 4385, which is lower than the initial solution, and therefore an improved solution. This procedure can be continued and possibly may improve the solution somewhat. It is, however, unlikely that much improvement can be obtained.

HEURISTIC PROBLEM SOLVING

What procedure was followed in the above solution procedure? The procedure followed is generally called a heuristic problem solving method. It allows improved solutions to be found by a trial and error approach which, nevertheless, has a definite amount of direction to it. For instance, the interchange of departments 3 and 5 was not arbitrary. A tool, the schematic diagram, represented by Figure 10-2 was used to derive information from the initial solution, which was not available otherwise. On the basis of the derived information action was taken which led to a better solution. Hence, one can see that the heuristic problem solving method is a powerful tool for making improvements to an initial solution. Unfortunately, one never knows whether the final solution obtained, or the solution obtained when the search for a better solution is stopped is the optimal solution.

One other disadvantage to the solution technique discussed above is the rather complicated procedure required to derive information regarding possible improvement. Granted the procedure is rather simple for the human brain if the problem is of small size, i.e., not complex. The above technique would not be very helpful for larger and more complex problems. However, if the procedure for the simple problem could be translated into a language which the computer could understand, then the computer's infinitely larger computing capacity could be used to tackle the larger and more complex problems.

Although the above procedure could theoretically be programmed for a computer, practically it would be such a huge and complicated project, making it economically infeasible to develop. However, there are other simpler heuristic rules which can be developed. These rules may not be very efficient on simple problems. However, since

the rules are simple and straightforward, these rules can
be incorporated in a computer program, and the computer's
large computing capacity can be used to help search for
improved solutions.

In the next chapter a heuristic procedure will be
presented and it will be shown that it can be a surprisingly
useful tool for the plant layout problem. The heuristic
methods will subsequently be followed by a combination
heuristic and optimization approach in the following chapter.

EXERCISES

1. Is the final solution for the example layout problem with a value of 4385 the minimum solution or is there a better layout?

2. Why is it important that constraints be imposed on the shapes of departments? Show that the value of the objective function (10-1) can be reduced to zero if no constraints are imposed on department configurations.

3. Ellicott Manufacturing does job machining and assembly and wishes to re-layout its production facilities so that the relative location of departments better reflects the average flow of parts through the plant. Below is shown an operation sequence summary for a sample of seven parts, with approximate area requirements for each of the thirteen machine or work centers. The numbers in the columns headed by each of the parts indicate to which work center number the part goes next. Just below the sequence summary is shown a summary of production per month and the number of pieces handled at one time through the shop for each part. Develop a load summary showing the number of loads per month going between all combinations of work centers.

Process Sequence Summary

Machine or Work Center	Area, Square Feet	Work Center Number	Part A	B	C	D	E	F	G
Saw	50	1		2	2				2
Centering	100	2		4	3				3
Milling machines	500	3	5	9	5	5		4	4
Lathes	600	4		5,7	7		5	10	5
Drills	300	5	8	3	4	11	7		6
Arbor press	100	6					11		7
Grinders	200	7		12	12		6		8
Shapers	200	8	9			3			9
Heat treat	150	9	11	4					10
Paint	100	10						11	11
Assembly bench	100	11	12	13	13	13	13	13	12
Inspection	50	12	13	11	11				13
Pack	100	13							

Production Summary

	A	B	C	D	E	F	G
Pieces per month	500	500	1600	1200	400	800	400
Pieces per load	2	100	40	40	100	100	2
Loads per month	250	5	40	30	4	8	200

4. Using the information contained in exercise 3 develop an ideal schematic layout using load summary information for Ellicott Manufacturing.

5. Develop a block diagram for Ellicott Manufacturing that reflects the approximate area requirements given and results in an overall rectangular shape. Use the data given in exercise 3.

6. Sanford Electrological is concerned with the large amount of material handling in its present plant. Based on the distance and load summary below determine if the present layout can be improved. Each department

occupies approximate equal amount of space and departments can
therefore be interchanged.

Distance Summary

	1	2	3	4	5
1	0	4	5	3	2.5
2	4	0	3	5	2.5
3	5	3	0	4	2.5
4	3	5	4	0	2.5
5	2.5	2.5	2.5	2.5	0

Load Summary

	1	2	3	4	5
1	0	60	100	40	100
2		0	60	60	0
3			0	0	60
4				0	30
5					0

7. Clearing Machine Works is in the process of moving to a new plant.
   The plant layout and methods department has laid out the plant in six
   approximately equally large departments and load calculations have
   determined the amount of material handling shown in the load summary
   below. The proposed layout will separate the six departments by
   distances as indicated on the distance summary shown below. Has the

plant layout and methods department developed the most efficient layout from a material handling point of view?

Distance Summary

	1	2	3	4	5	6
1	0	4	8	3	5	8.5
2	4	0	4	5	3	5
3	8	4	0	8.5	5	3
4	3.5	5	8.5	0	4	8
5	5	3	5	4	0	4
6	8.5	5	3	8	4	0

Load Summary

	1	2	3	4	5	6
1	0	60	0	40	10	80
2		0	40	0	20	0
3			0	0	100	20
4				0	0	80
5					0	20
6						0

8. Omega Electric is planning a new electric appliance plant. Only the non-electrical parts will be made at this new plant and the appliances will also be assembled. The space needs of the ten departments are as shown below

Department	Name	Space in Square Feet
1	Receiving	15,000
2	Blanking	6,000
3	Forming	12,000
4	Sub-assembly	9,000
5	Machining	45,000
6	Heat treat	24,000
7	Grinding	15,000
8	Painting	21,000
9	Assembly	63,000
10	Shipping	18,000
		228,000

The production engineering department had made an analysis of work flow between departments and had summarized this information in a load chart as shown below. The data in the load chart shows the number of loads to be transported between departments on a weekly basis for a given output rate. The cost of transporting each load can be assumed to be identical.

	1	2	3	4	5	6	7	8	9	10
1	0									
2	200	0								
3		200	0							
4	150		50	0			200	100		
5	200				0					
6					250	0				
7						300	0			
8	50		200	300				0		
9	500			200			150	200	0	
10	100		50	100			50	50	900	0

Develop an efficient layout for a one floor rectangular plant having internal dimensions of 600 feet long and 400 feet wide.

9. What additional information would you like to have prior to officially recommending the layout of the Omega Electric plant in exercise 8?

CHAPTER 11.--PLANT LAYOUT BY HEURISTIC METHODS

This chapter is a direct continuation of the previous chapter. The heuristic algorithm proposed by Armour and Buffa[1], and subsequently described in more detail by Buffa, Armour, and Vollman[2] will be reviewed and analyzed. The limitations of the heuristic technique are that only sub-optimum relative location or plant layout patterns are determined. The heuristic algorithm determines how plant layouts should be altered to obtain sequentially the most improved pattern with each change. The completed suboptimum plant layout is then presented as a block diagramatic layout of the facility areas.

The Buffa, Armour, and Vollman[3] article calls the heuristic approach CRAFT, which stands for "Computerized Relative Allocation of Facilities Technique." They report that the CRAFT program is in the SHARE[4] Library (No. SDA3391) of computer programs and is available for use by individuals or firms.

---

[1] G.C. Armour, and E.S. Buffa, "A Heuristic Algorithm and Computer Simulation Approach to the Relative Location of Facilities," Management Science, Vol. 9, No. 2 (January 1963), pp. 294-309.

[2] E.S. Buffa, G.C. Armour, and T.E. Vollman, "Allocating Facilities with CRAFT," Harvard Business Review, March-April 1964, pp. 136-58.

[3] Ibid., p. 136-58.

[4] The SHARE programs are available from IBM corporation or its subsidiary, Service Bureau Corporation.

This chapter will not describe in detail how the CRAFT computer program can be applied as the authors of the above references have adequately done. However, instead the above authors'proposed heuristic techniques will be described and illustrated with manageable illustrations. The descriptions and illustrations will be designed so that they form a bridge between the introductory material in the previous chapter and the more difficult material in the next and final chapter on plant layout.

THE MODEL

The objective as described in the previous chapter is to locate the departments or work facilities in such a way that material handling cost between departments is minimized. If it is assumed that the cost of moving a unit load over a unit distance is constant and if a standard piece of material handling equipment is used then minimizing material handling cost is equal to minimizing distance travelled.

The Objective Function

The two major components which make up the objective function consist of the number of unit loads, a_{jk}, moving between departments j and k in either direction. a_{jk} is an element of the hollow, real symmetric matrix A. The distance d_{jk}, between departments j and k can similarly be represented by the hollow, real symmetric matrix D.

If more than one kind of material handling equipment is used then the cost of moving material will vary. To accommodate the above let v_{jk} be the number of unit loads moving between departments j and k and let u_{jk} be the cost to move a unit load a unit distance between departments j and k.

Since all v_{jk} elements of the matrix V and all u_{jk} elements of matrix U are independent of departmental location, the product of v_{jk} and u_{jk} can be expressed in terms of a_{jk}, that is,

$$a_{jk} = v_{jk} \, u_{jk} .$$

Based on the above description of the matrix A the cost of moving a unit load from department j to k is the same as moving a unit load from k to j. The objective function to be minimized then becomes,

$$C = \sum_{j=1}^{n-1} \sum_{k=j+1}^{n} a_{jk} \, d_{jk}$$

where n is the number of departments; a_{jk} and d_{jk} are as defined above, j, k = 1, 2, ...,n; $a_{jk} = a_{kj}$; $d_{jk} = d_{kj}$;

$a_{jk} = 0$ for $j = k$; and $d_{jk} = 0$ for $j = k$.[5]

Combinatorial Problem

Assume that the n departments are considered as points to be assigned to n locations; then the number of layouts K is n!/M, where M, a small finite number is the appropriate measure of symmetry[6].

However, if the points are considered areas which they really are then K > n!/M, since as departmental configurations change the centers of departments and, hence, distances between centers change. Therefore, the combinatorial nature of the schematic relative location problem, which treats departments as points, is further compounded to an unknown extent by introducing consideration of area.

---

[5]If distances between departments differ when traveling in opposite directions, i.e. if $d_{jk} \neq d_{kj}$ for some j then the objective function, $C = \sum\limits_{j=1}^{n} \sum\limits_{j=1}^{n} a_{jk} d_{jk}$, should be used. The symbol a_{jk} then stands for the number of loads to be moved from j to k.

[6]There are fewer combinations than permutations. The adjacencies of departments are not altered if mirror image arrangements are made, or if the entire arrangement is rotated 180°, assuming rectangular shape. The measure of symmetry for rectangles is 4. Square arrangements may also be flipped about their diagonals. The measure of symmetry for squares is 8.

THE ALGORITHM[7]

It would be most desirable to derive a formal mathematical algorithm to minimize the objective function directly. However, in the event that the building of such an algorithm is not feasible, then a routine for arriving at location assignment patterns which are successively nearer the optimum is desirable. A heuristic algorithm to accomplish this is proposed below.

The algorithm consists of the following six steps.

1. Compute a matrix, D, of distances between computed department centers for the first feasible initial location pattern.

2. Compute the matrix, A, of number of loads between departments.

3. Evaluate the changes in C, ΔC, which would occur if each department was exchanged with all other departments in location wherever feasible. Find the largest ΔC. This requires a maximum of $(n^2 - n)/2$ evaluations. However, in evaluating the exchange of any two departments with each other it is only necessary to consider vectors headed by the two departments being evaluated for possible exchange. The evaluation process itself is, with one minor but essential modification, simply a process of computing two vector dot products, switching vectors, recomputing vector dot products and computing differences.

---

[7]Based on Armour and Buffa, op. cit., pp. 297-98.

To illustrate this, consider an exchange evaluation involving departments e and f. Pre-exchange partial transportation costs contributed by e and f are:

$$C_e + C_f = \sum_{k=1}^{n} d_{ek} a_{ek} + \sum_{k=1}^{n} d_{fk} a_{fk} - d_{ef}\, a_{ef}$$

Post-exchange partial transportation costs contributed by e and f are,

$$C'_f + C'_e = \sum_{k=1}^{n} d_{ek} a_{fk} + \sum_{k=1}^{n} d_{fk} a_{ek} - d_{ef} a_{ef}$$

Let ΔC_{ef} be the change in C resulting from an exchange of departments e and f. Then,

$$\Delta C_{ef} = \sum_{k=1}^{n} d_{ek} a_{ek} + \sum_{k=1}^{n} d_{fk} a_{fk} - \sum_{k=1}^{n} d_{ej} a_{fk} - \sum_{k=1}^{n} d_{fk} a_{ek} -$$

$$- 2 d_{ef} a_{ef}$$

Compare ΔCef to the sign and magnitude of the last ΔC_{jk} found. Retain the larger positive value. Continue until all exchanges have been evaluated.

4. If no positive ΔC_{jk} exists go to step 6. If a positive ΔC_{jk} exists make the exchange corresponding to the largest positive ΔC_{jk} found during step 3. Recompute D.

Print the new relative location pattern and associated cost
and move identifying information.

5. Go to step 3.

6. Stop. The sub-optimum has been reached.

AN EXAMPLE [8]

A small hypothetical problem will be used to illustrate
the procedure and intricacies of the problem. The hypothetical plant is rectangular, three hundred feet wide and
four hundred feet long. There are five departments to be
located. The areas of departments 1, 2, 3, 4 and 5 are
respectively 30000, 20000, 20000 , 20000 and 30000 square
feet. The number of loads to be transported between departments is shown by the upper triangular matrix A, (Table 11-1)
and the distances between departments in the original feasible
layout shown below is represented by the upper triangular
matrix D, (Table 11-2). Since it is only one of many layouts
to be evaluated each department is identified by a letter.
A feasible layout pattern is shown in Figure 11-1.

Number of Layouts

Since there are five departments there are a maximum
of $(n^2 - n)/2 = 10$ possible layouts. However, since there is
only a pairwise interchange between departments 1 and 2 and a three-

---

[8]In the next two examples only exchanges between equal-size
departments have been considered. A more thorough search would also
consider adjacent department interchanges and one for two departmen
interchanges.

Table 11-1

Number of Loads Matrix A

Departments	1	2	3	4	5
1	0	2	10	1	10
2		0	4	6	8
3			0	2	6
4				0	1
5					0

Table 11-2

Distance Matrix D

Departments	1	2	3	4	5
1	0	180	112	212	224
2		0	158	315	212
3			0	158	112
4				0	180
5					0

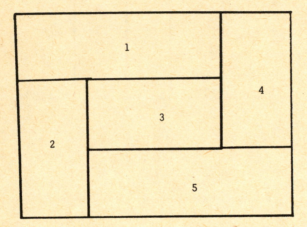

Figure 11-1

Feasible Layout Pattern

way interchange between 3, 4 and 5 the total possible layouts only number six. If the plant were assumed fixed
then there would be twelve possible layouts as shown in
Figure 11-2. Note, however, that from a material handling
point of view there are only six unique layouts. The
layouts are shown in a pairwise set in Figure 11-2 with the
respective cost values.

Assuming that the initial layout is the layout with
a cost of 9318 as shown in Figure 11-2 then the suboptimal
layout that would result from a complete iteration is the
layout with the cost value of 9034.

A Heuristic Solution

Inspecting the loads matrix A indicates that a large
number of loads is being transported in declining order
between departments 1 and 3, 1 and 5, 2 and 5, 2 and 4,
and 3 and 5. Taking this into account a good initial plant
layout appears to be the layout in Figure 11-3 determined
by graphic methods.

The cost of material handling for the layout in Figure
11-3 can be determined by first calculating the distances
between the departments. These are shown by matrix D in
Table 11-3.

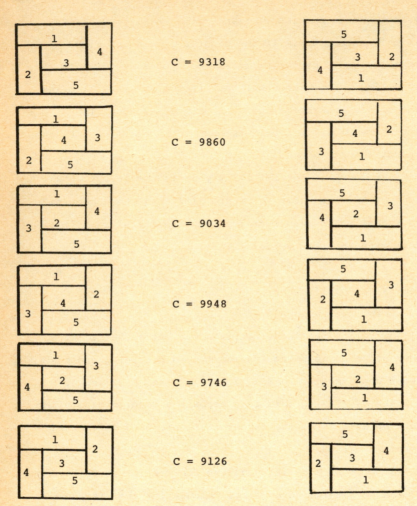

C = 9318

C = 9860

C = 9034

C = 9948

C = 9746

C = 9126

Figure 11-2

Feasible Layouts for One
Specific Pattern

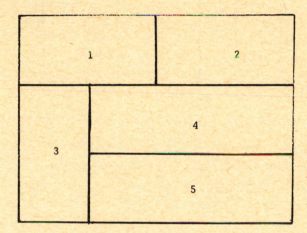

Figure 11-3

Graphically Determined Pattern

Table 11-3

Number of Loads Matrix A - Alternate Layout

Departments	1	2	3	4	5
1	0	200	158	180	250
2		0	292	112	206
3			0	206	206
4				0	100
5					0

The cost of material handling for the graphical layout is 9896 which is nearly as high as the worst possible layout on the previous set of layouts. Hence, what appears to be a good heuristic solution is not good at all but rather poor.

A SECOND EXAMPLE

To illustrate the pitfalls of heuristic sub-optimal solutions the following simplified example will be presented. The plant to be laid out is rectangular, is 200 feet wide and 300 feet long. There are four departments to be arranged rectangularly in multiple widths or lengths of 100 feet. Departments 1 and 2 cover 20000 square feet each and departments 3 and 4 cover 10000 square feet each.

The loads to be transported between departments are
shown in matrix A in Table 11-4.

Table 11-4

Loads Matrix A

Departments	1	2	3	4
1	0	5	7	4
2		0	2	1
3			0	9
4				0

The distance between departments is determined by calculating the distances between centers of departments
as was done in the previous example.

Performing the above calculations and fitting the departments onto the rectangular grid of 200 by 300 feet which
represents the plant results in eleven feasible layouts as
shown in Figure 11-4. The cost in material handling rounded
to hundreds of dollars is shown below each feasible layout and ranges from 3300 to 4700.

Using the algorithm previously presented, suboptimum
solutions are found which are frequently good but sometimes

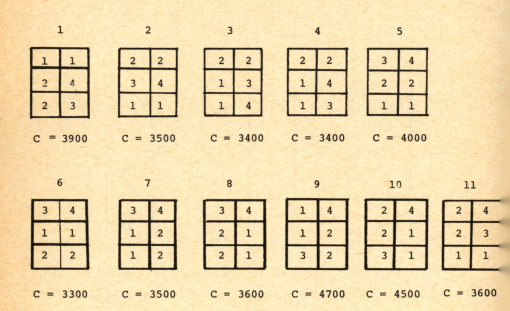

Figure 11-4

Feasible Plant Layouts

quite poor. For instance if the algorithm is started with layout 9 which has the highest cost of 4700 the suboptimum found is layout 10 with a cost of 4500. On the other hand if the starting layout were layout 5 with a cost of 4000 the suboptimum layout will be the absolute minimum cost layout 6 with a cost of 3300.

Continuing the above procedure will produce a suboptimum layout for each feasible layout. The absolute minimum cost layout is of course its own suboptimum. In Table 11-5 the results of this complete analysis are presented.

Table 11-5

Starting Layouts and Their Suboptimum Layouts

Starting Layout		Suboptimum Layout	
Number	Cost	Number	Cost
1	3900	3	3400
2	3500	2	3500
3	3400	3	3400
4	3400	4	3400
5	4000	6	3300
6	3300	6	3300
7	3500	7	3500
8	3600	7	3500
9	4700	10	4500
10	4500	10	4500
11	3600	3	3400

Table 11-5 thus reveals that if any one of the eleven feasible layouts was randomly selected there would be two chances in eleven of finding the optimum layout costing 3300; four chances in eleven of finding the next best layout costing 3400, three chances in eleven of finding a layout costing 3500 and two chances in eleven of finding the relatively poor layout costing 4500.

What the above analysis reveals is that the proposed heuristic method is a powerful tool to find good solutions. For instance the probability of finding a solution with a cost of 3500 or lower with two randomly-selected starting layouts is about 0.96 and with three randomly-selected starting layouts is about 0.99.

CONCLUSIONS

The technique discussed above is a powerful method to find good, optimum or near-optimum layouts. Without this aid a plant layout engineer has at his disposal only his qualitative judgments and schematic or graphic aids.

The use of the computerized technique discussed above assures that a superior layout is produced. This superior layout can be produced at a relatively low cost of computer time once the necessary data are available.

The above technique has been explained largely in terms

269

of plant layout. However, many other applications become
apparent especially in the case of multi-plant firms. The
methodology can be used for plant locations, selection of
a plant for a given department, warehouse locations and
so forth.

EXERCISES

1. Using the five-department layout in the first example,
find a good departmental pattern different from the two
patterns in the example. Determine the distance matrix D
and calculate the cost of material handling for your layout.

2. Find a good layout of a rectangular plant 200 feet wide
and 500 feet long. There are five departments of 20000, 30000,
10000, 20000 and 20000 square feet respectively. The loads
to be moved between departments are of two types: the first
type is transported at a cost of $0.075 per unit distance
and the second type has a transportation cost of $0.035 per
unit distance. The unit distance is 100 feet. The loads
matrix for the first type will be identified as V_1 and for
the second type as V_2. Both V_1 and V_2 are shown below.

	1	2	3	4	5
1	0	5	9	3	1
2		0	4	8	7
$V_1 = 3$			0	1	6
4				0	2
5					0

	1	2	3	4	5
1	0	1	8	3	6
2		0	3	1	8
$V_2 = 3$			0	7	3
4				0	4
5					0

Calculate the loads matrix A combining both V_1 and V_2. Calculate the distance between departments matrix D.

3. Find a good layout for problem 4. Use the heuristic algorithm with at least two starting layouts of different patterns.

4. Given the departments covering areas in hundreds of square feet as described below and the accompanying load summary, find a good layout using the SHARE computer program if available. The plant is of rectangular shape 200 feet wide and 300 feet long.

Load Summary

Department	Area	1	2	3	4	5	6	7
1	6000	0	20	10	30	25	60	10
2	2000		0	25	30	0	15	5
3	12000			0	10	15	0	35
4	4000				0	25	30	40
5	10000					0	10	15
6	10000						0	30
7	16000							0

5. Apply the heuristic method discussed in this chapter to exercise 6 in chapter 10. Assume that each department occupies 4000 square feet and the plant measures 200 by 100 feet. Ignore the distance summary.

6. Apply the heuristic method discussed in this chapter to exercise 7
    in chapter 10. Assume that each department occupies 5000 square
    feet and the plant measures 300 by 100 feet. Ignore the distance
    summary.

7. Sunset Electronics is in the process of setting up a new plant to
    produce electronic weapon detectors for use at airports to detect
    weapon carriers who may be potential hijackers. The plant consists
    of seven departments. Each department requires an area and has
    material handling requirements as shown below.

Department	Area	1	2	3	4	5	6	7
1	900	0	15	0	20	30	20	25
2	1500		0	40	10	25	40	65
3	1200			0	40	0	15	35
4	2100				0	10	25	20
5	2400					0	10	30
6	2700						0	40
7	1200							0

Find a good layout using the heuristic method described in this
chapter. What will the measurements of the rectangular plant be?

CHAPTER 12.--PLANT LAYOUT BY OPTIMIZATION METHODS[1]

In the previous chapter a heuristic method was presented to solve the relative location of facilities or
plant layout problem. In this chapter a combination of
heuristic and optimization technique will be reviewed.
Just because it is an optimization technique does not imply
that it is a better method than the one discussed in the
previous chapter. Whereas the method in the previous
chapter was kept simple so that it could be programmed
and executed relatively easily on a computer, the method
in this chapter is considerably more complicated and thus
far has not been programmed for an electronic computer.

The complex nature of the model presented in this
chapter has the advantage that the problem and solution
algorithm are shown in their full complexity. Hence, if
one wanted to computerize the algorithm, a number of simplifying assumptions could be introduced which would facilitate the programming project.

Before presenting the model a brief introduction of
the discrete optimizing technique used in the optimization
of the objective function will be provided.

----

[1]This chapter is based on C.C. Pegels: "Plant Layout
and Discrete Optimizing" The International Journal of
Production Research, Vol. 5, No. 1 (March 1966), pp. 81-92.

THEORY OF DISCRETE OPTIMIZING

The discrete optimizing technique is designed to attack problems such as that of the traveling salesman. The layout problem is a traveling salesman problem in which solution by complete enumeration would be extremely difficult. Complete enumeration becomes infeasible if the problem is very large. For instance the number of all possible routes of a twelve-city traveling salesman problem amounts to 11! or approximately 50,000,000 routes.

Furthermore, complete enumeration ignores the cost of computing, which, in large problems, becomes formidable. A solution derived from complete enumeration with computing cost ignored is, therefore, not optimal in the strictest sense. The discrete optimizing technique developed by Reiter and Sherman[2] uses statistical sampling to escape the need for complete enumeration and considers computing cost when deciding to terminate sampling.

Discrete Optimizing Procedure

The procedure is as follows: A sample layout is drawn at random from all possible layouts, and called an element p from the set of all possible layouts P, (pϵP). The randomly chosen layout p is evaluated in terms of cost

---

[2]S. Reiter and G.R. Sherman, "Discrete Optimizing," SIAM Journal, Vol. 13, No. 3 (1965), pp. 864-89.

or some other measure. The value of the layout is then
denoted as $f(p)$. Let f be rounded to the nearest integer
and have as upper bound the integer f_u and as lower bound
the integer f_ℓ. Let Z be the set of ordered integers $Z(i)$,
$i = 1,..., R$ such that $Z(1) = f_u$ and $Z(R) = f_\ell$, then $f(p)\epsilon Z$
for all $p\epsilon P$.

Calculation of Local Minimum

The sample layout is modified according to a specified rule, producing another layout, which is evaluated.
If the new layout is better than the previous layout, it
is retained, and if it is worse, it is discarded. This
search procedure is repeated for the complete neighborhood
of the original sample layout. It should be noted that the
rule for modifying an initial layout defines the neighborhood.[3] The best layout in this neighborhood then becomes
the locally minimal layout of that neighborhood. The value
of this local minimum is denoted M(1).

The local minimum corresponding to the first sample
having been found, a second sample layout is drawn. This
produces another local minimum with value M(2), which may
or may not equal M(1). Successive draws from the set of all
possible layouts generate a sample from the population of

---

[3]For a detailed description of neighborhoods see Reiter
and Sherman, op.cit., pp. 864-89.

local minima, M(1), M(2), M(3),.... The generated probability distribution of values of local minima is then used to estimate the improvement possible from further search.

Development of Tree Structure

If each layout in P were evaluated as above, then P would be factored into exhaustive and mutually exclusive subsets. These subsets can be pictured in the form of trees with layouts p at the nodes of the trees and a minimum layout at the bottom of the tree. Since f may take on the same value at different points, the minima do not necessarily have unique values. In other words, the set Z may contain elements that are minima for several trees and elements that are not minima for any trees.

The above procedure can be illustrated by the tree structures shown in Figure 12-1. The set P consists of ten layouts, and there are four minima. Let

$$Z = \{Z(1),Z(2),Z(3),Z(4),Z(5)\} = \{6,5,4,3,2\}.$$

If the layouts in P are sampled at random, it can be said that the layouts are uniformly distributed. The sampling determines a probability distribution U over Z. U is a multinominal distribution, and according to the law of probability $\Sigma U = 1$. Hence $U(1) = 0.1$, $U(2) = 0.4$, $U(3) = 0$, $U(4) = 0.2$ and $U(5) = 0.3$.

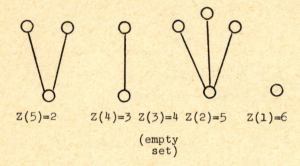

$Z(5)=2$ $Z(4)=3$ $Z(3)=4$ $Z(2)=5$ $Z(1)=6$

(empty
set)

Figure 12-1

Tree Structure

The objective of discrete optimizing is to find that layout such that the average cost of finding a better p exceeds the expected return from finding that better p. A stopping rule is thus found which determines when the search ends.

DISCRETE OPTIMIZING AND THE PLANT LAYOUT PROBLEM

The objective, as mentioned in the introduction, is to minimize cost. Two kinds of cost can be incurred, transportation cost, V_1, and penalty cost for breaking up departments into sections, V_2.

Definition of the Problem

To simplify the problem, the plant is assumed to be a rectangular grid, although, from a theoretical point of view, any configuration of plant is feasible. The length L, with grid index $\ell = 1,2,\ldots, L$, and width W, with grid index $w = 1,2,\ldots, W$ determine the plant's area, LW. If L and W are measured in appropriate units, the area LW will be an integer and factorable in several L-W combinations.

Each department x, where $x = 1,2,\ldots, X$, may be broken up into sections of certain specified sizes. For instance, a department occupying 100 cells might be broken into four sections of size 40, 20, 20 and 20. The section sizes are restricted by the manufacturing process. A vector b represents the number of cells occupied by each department section, that is, department section j occupies b_j cells. Hence

$$\sum_{j=1}^{n} b_j = LW$$

Development of Transportation Cost Component

The number of loads to be transported between department sections in either direction can be represented by a hollow, real, symmetric matrix, A. ($A = \| a_{jk} \|$, $j,k = 1,2,\ldots, n$; $a_{jk} = a_{kj}$; and $a_{jk} = 0$ for $j = k$.) The distances between department sections can be represented by a hollow, real, symmetric matrix, D. ($D = \| d_{jk} \|$ $j,k, = 1,2,\ldots,n$; $d_{jk} = 0$ for $j = k$; and $d_{jk} = d_{kj}$.) The distances can be defined as,

$$d_{jk} = \left[(\bar{\ell}_j - \bar{\ell}_k)^2 + (\bar{w}_j - \bar{w}_k)^2 \right]^{\frac{1}{2}}, \qquad (12\text{-}1)$$

where $\bar{\ell}_j, \bar{\ell}_k, \bar{w}_j$ and $\bar{w}_k$ are the average values of ℓ and w for the department sections j and k.

The present value of transportation cost will be defined as:

$$V_1 = v_1 \beta \sum_{j=1}^{n-1} \sum_{k=j+1}^{n} a_{jk} d_{jk},$$

where v_1, is constant cost per load-distance and

$$\beta = (1-e^{-rt})/r$$

which is the present value factor. For derivation of this formula see Davidson, Smith and Wiley.[4]

The present value factor changes a flow of cost over a future time period of length T into an equivalent current cost if capital cost or interest r are known. It should also be noted that "e" is a constant equal to 2.7183. As an example it will be assumed that the useful life of a planned production facility is 120 months and capital cost is 9 percent annually, 0.75 percent per month, then T = 120 and r = 0.0075. The present value factor then becomes

$$(1-e^{-0.90})/0.0075 = 79.124,$$

and assuming that v_1 = \$0.10 and that the total number of loads to be moved times the distances per respective moves,

$$\sum_{j=1}^{n-1} \sum_{k=j+1}^{n} a_{jk} d_{jk},$$ amount to 100,000 units per month, then

the equivalent current or present value cost, V_1, can be calculated as follows:

---

[4] R.K. Davidson, V.L. Smith, and J.W. Wiley, Economics: An Analytical Approach, Homewood, Illinois: R.D. Irwin, Inc., 1962, p. 116.

$$V_1 = (79.124)(0.10)(100,000) = \$791,240.$$

Penalty Cost for Departmental Split Ups

The penalty cost for departmental split-up into sections is $v_2(N_x-1)$ per split-up, where N_x is the number of sections into which department x is split and

$$V_2 = v_2\beta \sum_{x=1}^{X} (N_x-1)$$

is the present value of the total penalty cost for split-ups.

Assuming that v_2 is estimated to be \$50 per split-up, if department x is split into three sections (equivalent to two split-ups) then the split-up cost for department x will be $v_2(N_x-1) = (50)(2) = \$100$ per month. If the total monthly split-up cost for all departments is \$1000, then this monthly cost flow can again be changed to an equivalent current cost by the present value factor. Total present value split-up cost thus becomes

$$V_2 = (79.124)(1000) = \$79,124.$$

Summary of Two Cost Items

Let P be the set of all possible layouts of the department sections. Hence, $p \epsilon P$ is a layout, and the total cost per layout is $f(p) = V_1+V_2 = \$870,364$. If computing cost is ignored, an optimal layout, p^*, has been found when $f(p^*) \leq f(p)$ for all $p \epsilon P$.

The proposed algorithm is only one of many that may be used, provided the basic technique of discrete optimizing is adhered to. The choice of an algorithm is an optimizing problem by itself, since small neighborhoods with low computing cost per neighborhood provide a smaller probability of finding the absolute optimum than do larger neighborhoods with higher computing cost. The choice of an algorithm really entails a decision about the size of the neighborhoods to be employed.

THE ALGORITHM

The algorithm consists of the following steps:

1. Determine the unit of measurement for L and W, which in turn determines the size of the cells in the grid. (The size of the cells in the grid is determined by the unit of measurement in this algorithm and in the hypothetical example. Although it may complicate the problem somewhat, it is possible to make the cells either larger or smaller than the unit of measurement.) If L and W are measured in yards, then the area of a cell is one square yard.

2. Randomly choose L to be an integer in the interval, $0 < L < LW$ (LW is area). W will then be an

integer, W = LW/L.

3. Randomly determine the number of sections into which each department will be split. The number of sections for department x is N_x such that N_x satisfies the imposed constraints. The total number of sections in the plant will then be,

$$n = \sum_{x=1}^{X} N_x.$$

4. Compute the vector b and the matrix $A = \| a_{jk} \|$, $j,k = 1,2,\ldots,n$.

5. Randomly select an (ℓ,w) from grid LW. Select the largest section j, from vector b. (The largest section with lowest index should be selected where sections are otherwise equal.) Enter b_j cells of section j onto grid LW in an appropriate sequence--i.e. ℓ; w; ℓ, w-1; ℓ, w-2-- continuing until all b_j cells have been entered onto grid LW. Again randomly select an (ℓ,w) and repeat the above procedure for the next largest section until all department sections have been positioned onto grid LW. Denote this initial arrangement p_1.

6. At this point the elements of the matrix D can be calculated by formula (12-1).

7. Calculate $f(p_1)$ for formula $f(p) = V_1 + V_2$.

8. Interchange sections that are adjacent to each other, or sections that are of equal size in terms of numerical area units in a specified sequence. A proposed order of interchange for a plant with four sections is the sequence, 1-2, 1-3, 1-4, 2-3, 2-4 and 3-4. If any of the interchanges is not possible it is passed over. The matrix D is determined, and $f(p)$ is calculated for each interchange. The layout with the lowest $f(p)$ is retained and compared with each succeeding layout. The minimum cost layout, or in the case of equality, the first minimum cost layout, is denoted p_2. If none of the layouts provides a lower cost than p_1, then the iteration is stopped and p_1 is the local minimum layout of its neighborhood.

9. If $f(p_2) < f(p_1)$ then step 8 is repeated, starting with layout p_2. If a lower cost layout is found, after completing all possible interchanges, it is denoted p_3. This procedure is continued until a complete repetition of step 8 is made without improvement. The lowest cost layout found is then the local minimum layout of the neighborhood analyzed.

THE STOPPING RULE

The next problem facing the decision maker is when to stop the sampling procedure. Just one complete neighborhood search of a realistic problem may be of considerable size even for the larger computers. A true optimum is reached only if computing cost is taken into account. Hence the sampling procedure is stopped when the average cost of locating another local optimum exceeds the return from locating a better local optimum times the probability of locating that better local optimum. The stopping rule will now be discussed.

Probability of Having Found Optimum Layout

Random sampling from the population of all possible layouts, prior to each neighborhood search, does not provide any certainty that the optimum is ever reached. However, as in all statistical investigations based on stated assumptions, there is a probability of $1-\alpha$, that the optimum is reached. The error probability, α, is the probability that the absolute minimum has not been found if the computation is stopped at the orders of the stopping rule.

The stopping rule provides information on how many additional samples should be drawn and evaluated after attaining an improvement over the previous "best" layout. For example, if the best layout is p with cost $f(p)$, the stopping rule can be used to calculate the number, K, of additional samples to draw before stopping. If prior to the Kth sample a layout p' with cost $f(p')$ is found such that $f(p') < f(p)$ then K can be re-calculated before proceeding.

Stopping Rule Formula

The stopping rule is of the following form. (The mathematical development of this rule is given in the Appendix.)

$$K = \frac{m\ln\alpha + \dfrac{mC}{M^*(m)-Z(R)}}{m\ln\alpha - \dfrac{mC}{M^*(m)-Z(R)}}$$

where

K = number of additional samples to be drawn, and neighborhoods to be evaluated, before stopping;

m = number of neighborhoods evaluated thus far;

α = error probability;

C = average cost of evaluating a neighborhood;

$M^*(m)$ = the cost of the "best" local minimum located thus far;

$Z(R)$ = the absolute minimum value. Although this value may be known or can be closely estimated, the layout by which it is determined is not known.

How to Apply Stopping Rule

To use the above stopping rule, the decision maker must specify α, calculate C and estimate Z(R). The latter two can only be determined after the computation has started. In some combinatorial type problems the value for Z(R) is known; in the plant layout problem it is not known, and may never be known, with certainty. Hence the proposed alternative is to estimate Z(R) from the generated locally minimal points M(1), M(2), M(3),..., and revise Z(R) as better points are generated. (One rule for estimating Z(R) would be Z(R) = $M^*(m) - \delta(m)$, where $\delta(m) > C$. $\delta(m)$ should be a decreasing function of m and should increase as α increases. A reasonable formula for $\delta(m)$ seems to be,

$$\delta(m) = C(1+\alpha)(m-1)/m + \delta(1)/m^{1/\gamma}$$

where $\delta(1)$ is estimated after calculating the first local minimum and γ, $\gamma > 1$, is specified by the decision maker.

AN EXAMPLE

The hypothetical plant mentioned in the introduction is used to illustrate the algorithm. The required area, LW, is six units. The described algorithm is slightly modified to keep this example to manageable size for hand computation. This modification entails doing steps 1,2,3

and 4 only once instead of repeatedly. The length L = 3
and the width W = 2 units. Table 12-1 shows the possible
divisions of the three departments into sections. Department 1 can be broken up into three or two sections, and
Department 2 into two sections.

Table 12-1

Minimum Size of Sections of Departments

Department	Size of sections			Department area
1	1	1	1	3
2	1	1	0	2
3	1	0	0	1

Random selection determines that $N_1 = 2$ and $N_2 = 1$.
The new size of the department sections is shown in Table
12-2.

Table 12-2

Determined Size of Sections of Departments

Department	Size of sections		Department area
1	2	1	3
2	2	0	2
3	1	0	1

In order to simplify the following analysis the
sections are renumbered as follows: section 11 to 1,
section 12 to 2, section 21 to 3 and section 31 to 4.

The vector $b = (b_1, b_2, b_3, b_4) = (2, 1, 2, 1)$. The matrix $A = \| a_{jk} \|$, $j, k = 1,2,3,4$ has the following form,

$$A = \begin{bmatrix} 0 & 5 & 7 & 4 \\ 5 & 0 & 2 & 1 \\ 7 & 2 & 0 & 9 \\ 4 & 1 & 9 & 0 \end{bmatrix}$$

Figure 12-2 shows that there are eleven possible layouts of the plant. For each layout the matrix D can be computed. Letting $v_1 = r/(1-e^{-rt})$ and $V_2 = 0$, $f(p)$ is calculated for each layout by the formula,

$$f(p) = \sum_{j=1}^{3} \sum_{k=j+1}^{4} a_{jk} d_{jk}.$$

In Figure 12-2 each layout is numbered from 1 to 11, and the cost of each layout, $f(p)$, is shown directly below it.

Equivalent Layouts

Before proceeding with the problem the reader should be aware that there are a number of equivalent layouts that may be generated. For instance the layout

11
22
34

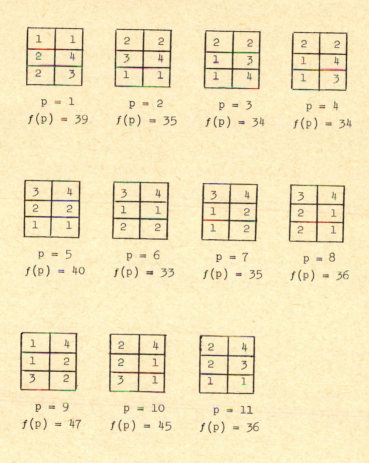

Figure 12-2

Feasible Plant Layouts

is equivalent to

```
        11      34          43
        22, to 22 and to 22.
        43      11          11
```

It may be argued that the above layouts are not really equivalent but identical. Assuming that they are equivalent, each original layout will produce a neighborhood and a local minimum that is equivalent to neighborhoods and local minima produced by equivalent original layouts.

Applying the Algorithm

Applying the algorithm, that is steps 5, 6, 7, 8 and 9 to the problem produces the previously discussed tree form as shown in Figure 12-3 with layouts at the node of the tree and local minima at the bottom of the tree.

It will be seen from Figure 12-3 that there are four minimal layouts, of which two have the same cost value. Each minimal layout is a local minimum for one or more neighborhoods.

Assuming that the layouts are uniformly distributed, Figure 12-3 indicates that the absolute minimum cost of 33 will occur with a probability of 5/11, and the costs of 34 and 35 with probabilities of 3/11 each. Hence the probability of not hitting the absolute minimum in five attempts is less than 0.05. (Based on sampling with replacement and

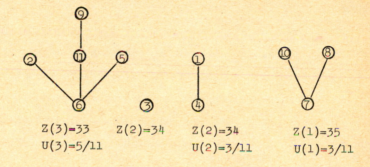

Figure 12-3

Tree Structure of Plant Layout

consequent use of binomial formula:

$$(5/11)^0 \ (6/11)^5 \doteq 0.045.$$

Application of Stopping Rule

An application of the stopping rule will now be presented. It will be assumed that three neighborhoods have been evaluated, that the lowest local minimum cost found thus far is 34, that the error probability is set at 0.05, and that the average cost to evaluate a neighborhood is 0.25 in equivalent cost units. The number of additional neighborhoods which should be evaluated before stopping by the stopping rule formula can then be determined. Applying the formula, it is found that $K = 2.53$ which is rounded to 3. Three additional neighborhoods should thus be evaluated before stopping the iteration. With this simple problem it appears extremely likely that the absolute minimum will be found.

Final Comments

The above technique applied to a simple problem seems, and is, rather redundant. Applied to realistic problems, however, it becomes a powerful tool to solve combinatorial type problems like the plant layout problem discussed in this chapter.

The method presented solves the plant layout problem when the cost of material handling is the criterion. Since computing cost is considered, it is a method that will provide an optimal or near optimal solution.

The proposed algorithm is complex and requires considerable computing time. However, plant layouts of whole plants are not daily decisions, and the potential return of an improved plant layout may well be worth the investment in programming and computer time.

EXERCISES

1. Given the tree structure below, what is the probability distribution U? What is the minimum value of Z(j) for which $\sum_{i=1}^{j} U(i) = 0.60$?

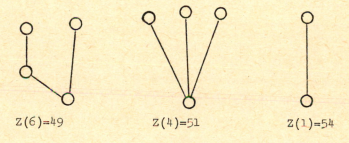

Z(6)=49   Z(4)=51   Z(1)=54

2. Suppose discrete optimizing is applied to a traveling salesman problem. A salesman lives in city A and visits cities B, C, D and E. He wants to minimize the cost and

inconvenience of traveling. Traveling costs including inconvenience costs between cities are itemized below.

	A	B	C	D	E
A	0	5	9	27	16
B	5	0	13	32	15
C	9	13	0	19	23
D	27	32	19	0	14
E	16	15	23	14	0

Find the minimum cost route visiting all cities from
A to A by heuristic methods.

3. Using the cost data in problem 2 develop a procedure for
   finding an initial solution.

4. Using the initial solution in problem 3 develop a rule
   for modifying the layout so that a neighborhood is defined. What is the local minimum cost of your first
   neighborhood?

5. Develop a tree structure of your problem and from the
   tree structure develop a probability distribution U.

6. Repeat the plant layout example problem with four departments of size 3, 6, 2 and 9. Assume that all
   departments must be rectangular and that no departmental
   split ups are allowed.

7. In what other areas could the discrete optimizing techniques be used?

8. Smith Manufacturing is planning a new plant with five departments of
5000, 6000, 8000, 9000, and 4000 square feet respectively. The loads
to be moved between departments are shown in the load summary below

Load Summary

	1	2	3	4	5
1	0	70	120	130	20
2		0	50	100	10
3			0	40	30
4				0	140
5					0

Apply the discrete optimizing technique to solve for the optimal plant
layout for Smith.

9. Apply the discrete optimizing technique to solve the plant layout problem
for Clearing Machine Works (exercise 7 of chapter 10). Assume that
each department occupies 5000 square feet and the plant is rectangular
and measures 300 by 100 feet.

APPENDIX

Let $M^*(m)$ be the minimum of the m observations, i.e.

$$M^*(m) = \text{Minimum } \{M(1), M(2),\ldots, M(m)\}.$$

The general stopping rule then states, continue sampling until,

$$\sum_{i > i_m} U(i)[M^*(m)-Z(i)] < C$$

where C is the average cost of computing a local minimum and where i_m is the i for which $Z(i) = M^*(m)$.

$Z(R)$ is the minimum over the set Z. Since sampling should be stopped when,

$$\sum_{i > i_m} U(i)[M^*(m)-Z(i)] < C,$$

it should certainly stop when,

$$\sum_{i > i_m} U(i)[M^*(m)-Z(R)] \leq C.$$

Both $M^*(m)$ and $Z(R)$ are independent of i, and $M^*(m) \geq Z(R)$. Therefore the stopping rule can be written,

$$\sum_{i > i_m} U(i) \leq \frac{C}{M^*(m)-Z(R)}$$

Observe that

$$\sum_{i>i_m} U(i)$$

is the probability of an improvement on the next sample draw.

A sufficiently long run of observations without improvement can be regarded as evidence that the probability of improvement is small. Assume that sampling is stopped after a run of K observations without improvement. Before the run, the supposition is that the probability of an improvement on the next draw, θ, has the density,

$$Pr(\theta) = 1 \text{ for } 0 \leq \theta \leq 1.$$

It is reasonable to risk with probability α that the termination is premature, i.e.

$$Pr\left(\theta > \frac{C}{M^*(m)-Z(R)}/K\right) = \alpha .$$

The probability of the run given θ is

$$Pr(K/\theta) = (1-\theta)^K$$

so the marginal probability of the run is

$$Pr(K) = \int_0^1 (1-\theta)^K d\theta = \frac{1}{K+1}.$$

The conditional density of θ given the run is,

$$Pr(\theta/K) = (K+1)(1-\theta)^K \quad \text{for } 0 \le \theta \le 1,$$

and the desired error probability is,

$$Pr(\theta > \gamma/K) = \int_\gamma^1 (K+1)(1-\theta)^K d\theta = (1-\gamma)^{K+1}$$

where $\gamma = \dfrac{C}{M^*(m)-Z(R)}$

Thus, one should stop after K observations without improvement where,

$$K = \frac{\ln \alpha}{\ln \left[1 - \dfrac{C}{M^*(m)-Z(R)} \right]} - 1$$

One feature of this termination rule is that the last K observations do not improve the solution. The cost of these last K observations is incurred to demonstrate that a near-optimum point has been found, and not to improve the solution. This means that the effective cost per local minimum is higher than the calculating cost C. In particular, the effective cost per local minimum is mC/(m-K). Inserted in the preceding formula,

$$K = \frac{\ln \alpha}{\ln \left(1 - \dfrac{mC}{\{M^*(m)-Z(R)\}(m-K)} \right)} - 1$$

this correction implies (based on ln (1-x) $\doteq$ -x) that,

$$K \doteq \frac{m \ln \alpha + \dfrac{mC}{M^*(m)-Z(R)}}{\ln \alpha - \dfrac{mC}{M^*(m)-Z(R)}}.$$

PART IV.   OPERATIONS PLANNING MODELS

Operations planning models will be explored in the last six chapters. Facility planning is so dependent on operations planning that a treatise such as this book would be incomplete without at least a review of some of the operations planning models. The models discussed in this last part therefore complement the previously-discussed facilities planning models.

In Chapter 13 the effects of learning will be discussed. Especially in launching new facilities or in launching a new product the cost of the start up phase is considerable. Considerable work has been done in the past to understand how learning takes place and how productivity increases over time. The classical techniques will be described and analyzed and a new learning model will be introduced which overcomes some of the disadvantages of the classical model.

Learning curves is one management tool that has been around for many years. Even prior to World War II it was recognized in the aircraft industry that the cost of producing the first few airplanes of a particular model was much more expensive than subsequent units.

Presently, the learning curve is used by the defense department to estimate the costs of major armaments production as a basis for negotiating prices. It is also used extensively by subcontractors to determine the costs associated with the completion of subcontract bids. The smaller the number of units to be produced the higher the unit cost will be. Hence, for cost estimating in bidding on contracts a familiarity with the concept of learning is essential.

Other industries where learning curves are or should be used are in electronic products assembly, residential home construction, shipbuilding and machine shops. In all these areas there is sufficient but limited volume of a given model. Hence learning takes place and should be considered in cost estimating and planning.

Anthony[1] presents a financial analysis case entitled "Lacklin Aircraft" where the purchasing agent of an aircraft firm uses the learning curve to obtain a reasonable but low price from a subcontractor for supplying aircraft subassemblies. Since the purchasing agent was familiar with the learning curve he had much better knowledge of the actual cost structure of the subcontractor than the subcontractor thought he had.

Chapters 14 and 15 are concerned with demand forecasting. In Chapter 14 demand forecasting on the basis of external factors is discussed and models are presented which can be used for forecasting over the longer term. Chapter 15 discusses mechanical forecasting methods such as moving averages and exponential smoothing. These forecasting methods are applicable to those cases where demand forecasts for a large number of items must be made frequently and at low cost. Forecasting methods employed only depend on historical demand and ignore external factors. Where external factors affect demand the mechanical forecasting methods should of course be avoided.

Forecasting based on exogenous factors as presented in Chapter 14 is probably the most common method of forecasting demand. Each firm must go through some sort of forecasting process before it can prepare its budgets for the coming year. Whatever the process used a forecast is produced. The quality (or accuracy) of the forecast is usually a function of the forecasting process

---

[1] Anthony, R.N., _Management Accounting_, Homewood, Illinois: Richard D. Irwin, Inc., Fourth Edition, 1970.

Parker and Sequra[2] report that for many firms regression analysis as reported in Chapter 14 can predict more accurately than less scientific methods can. They report that The American Can Company for several years has used a regression technique to estimate sales on the basis of numerous external factors. They cite the example of how this firm forecasts beer-can demand by using an equation that correlates sales to income levels, number of drinking establishments per thousand persons, and age distribution of the population. They cite other well known firms such as Eli Lilly, RCA Sales Corporation and Armour who have used mathematical techniques to forecast sales of their products. They conclude their article by pointing out that forecasting remains an imprecise art and subjectivity and clear judgment will continue to play a crucial role in it as long as the future refuses to replicate the past.

Chambers, Mullick and Smith[3] report on forecasting models used by Corning Glass Works. They point out three basic methods of forecasting. These are: qualitative models such as opinion polls, committee forecasts, and so forth, discussed in Chapter 14; time series analysis and projection such as is used in exponential smoothing discussed in Chapter 15; and causal models used in regression analysis as briefly described in Chapter 14. They conclude their article with the advice to the developers of forecast models to keep the knowledge and experience of managers in mind.

---

[2] Parker, G. C. and E. L. Segura, "How to Get a Better Forecast," Harvard Business Review, March-April 1971, Vol. 49-2, p. 99.

[3] Chambers, J. C., S. K. Mullick and D. D. Smith, "How to Choose the Right Forecasting Technique," Harvard Business Review, July-August 1971, p. 45.

Exponential smoothing methods are widely used in industry and in non-industrial situations. The author knows of several multiproduct firms that use exponential smoothing for forecasting sales levels of their products. Since it is a rather common and accepted method of forecasting demand of individual items little has been published on successful applications. An interesting application of this technique to blood demand forecasting, or non-industrial application, is described in a paper by Frankfurter, et al.[4]

The last three chapters, Chapters 16, 17 and 18 are concerned with planning work force, production and inventory. Since a planned production facility must accommodate the work force and frequently also the inventory, an understanding of the interaction between work force, production and inventory is important. Chapter 16 discusses two models which derive linear decision rules for production planning. The first model is for a plant which can build up inventory to smooth production levels and as a result linear decision rules are derived for both production and work force levels. The second model is for a job shop operation where inventory is not feasible and the linear decision rule derived is for work force level only.

Holt, et al[5] report on the application of decision rules for production, inventory and work force levels to a small paint manufacturing plant. They discovered that total costs could be reduced by over 7.5 percent. An empirical test revealed that fluctuations in aggregate production, back orders

---

[4]Frankfurter, G. M., K. E. Kendall, C. C. Pegels and P. D. Wharton, "Management Control of Blood through a Short Term Supply-Demand Forecast System," School of Management, 1971

[5]Holt, C. C., F. Modigliani, J. F. Muth and H. A. Simon, Planning Production, Inventories and Work Force, Englewood Cliffs, N.J.: Prentice-Hall, Inc., 1960.

and inventory levels were reduced also. Following the empirical test

management adopted the decision rules as guides in production planning. An

application by the author described in Chapter 16 to a prefabricated home

builder indicated that a modified decision rule model could result in

substantial savings. Unfortunately, the model was never applied because of

instability in the prefabricated home market at the time.

Chapter 17 uses a linear programming approach to determine work force

and production levels for a plant. It is somewhat easier to apply because

linear cost functions are used instead of the quadratic cost functions, based on

linear cost functions, used for the models in Chapter 16. Whether the model

provides better results is of course dependent on how cost behaves, i.e., is cost

linear or non linear.

This model was developed following the work by Holt et al discussed in the

previous chapter. The author is not aware of any experiences with this linear

programming technique in applied situations. Some work has been done on compar-

ing the linear programming model with the quadratic cost function model. However,

no hard conclusions could be drawn from this unpublished study.

Chapter 18 uses a novel approach to work force and production planning

over the somewhat more distant horizon. It uses the technique of input-output

analysis developed for macro-economic analysis. The model is essentially a

production planning model which determines the production levels for the produc-

tion departments of a plant which produces assembled products and component

parts of the assembled products. Based on the required production plans the

work force level can then be determined. The method was developed for a

machinery manufacturer and is based on an input-output model of the firm's production and inventory system. It has been used for a number of years as a basic tool in the preparation of the company's operating plans as well as for a variety of special studies.

CHAPTER 13.--PROJECTION OF PRODUCTION COST DURING START UP

PHASE[1]

This chapter will review what has been developed and
used to estimate and project future costs of a new product
or of a new production facility. In addition a new model
called the exponential function will be described and compared with the traditional power function formulation.

INTRODUCTION

Learning curves, startup curves,   cost reduction
curves, or by whatever other name they are known, have
many potential applications in the areas of production planning, budgeting, purchasing, and contract bid preparation.
A review of learning curves applications by Baloff[2] indicates that the power function formulation can be used to
describe the productivity increases that accompany the intro-
ductions of new products in a variety of labor-intensive
assembly situations, including the manufacture of airframes,
electronic and electro-mechanical components, and machine
tools. In addition, the model has also found applicability
to instances of both product and process startup in the

---

[1]This chapter is based on C. Carl Pegels: "On Startup
or Learning Curves: An Expanded View," AIIE Transactions,
Vol. 1, No. 2 (September 1969), pp. 216-22.

[2]Nicholas Baloff, "Estimating the Parameters of the Start
Up Model-An Empirical Approach,"The Journal of Industrial Engineering
Vol. 18, No. 4 (April 1967), pp. 248-53.

machine-intensive manufacture of steel, glass, paper, and electrical products.

Standardized Power Function

The literature so far has concentrated mainly on a somewhat standardized power function formulation. This formulation was first introduced by Wright[3] in 1936. Since then this power function formulation has managed to survive in essentially its original form, even though it has many basic weaknesses.

Alternative Learning Functions

However, a number of alternative functions have been proposed. Most of these were intended for specific applications and therefore have not affected the popularity of the power function to any degree. For instance Carr[4] proposed an S-type function which was based on the assumption of a

---

[3]T. P. Wright, "Factors Affecting the Cost of Airplanes," Journal of the Aeronautical Sciences, Vol. 3, February 1936, pp. 122-28.

[4]G.W. Carr, "Peacetime Cost Estimating Requires New Learning Curves," Aviation, Vol. 45, April 1946, pp. 76-77.

gradual start-up. Guibert[5] proposed a complicated multi-
parameter function with several restrictive assumptions.

More recent work holds somewhat more promise. Among
these, DeJong[6] proposed a version of the power function which
generates two components, a fixed component which is set
equal to the irreducible portion of the task, and a variable
component, which is subject to learning. Levy[7] has presented
a new type of firm learning function and shows it to be useful in explaining how firms adapt to new processes, and in
isolating the variables that may influence the firm's rate
of learning. Levy's learning function reaches a plateau and
does not continue to decrease or increase as does the power
function. Asher[8] reports on a variety of different approaches,
which were proposed mostly during and immediately following
World War II. The main drawback of these proposed functions
is the difficulty associated with parameter estimation. However, all the proposed alternatives have not been able to
dislodge the power function which is apparently still the
most common one in use at the present time. It is, of course,

---

[5] P. Guibert, Le Plan de Fabrication Aeronautique, Paris:
Dunod, 1945. English translation under the title, Mathematical
Studies of Aircraft Construction is available from Central Air
Documents Office, Wright-Patterson Air Force Base, Dayton, Ohio.

[6] J.R. DeJong, "The Effects of Increasing Skills on Cycle
Time and Its Consequences for Time Standards," Ergonomics,
Vol. 1, No. 1 (1957), pp. 51-59.

[7] F.K. Levy, "Adaption in the Production Process," Management Science, Vol. 11, No. 6 (April 1965), pp. 136-54.

[8] Harold Asher. Cost-Quantity Relationship in the Airframe Industry, Santa Monica, California: The Rand Corporation, 1956.

possible that the now-common power function formulation is the "best" algebraic function to describe "learning" or "productivity" increases, but on the other hand, this same power function also has some disadvantages.

Proposed Exponential Function

An alternative algebraic function, an exponential type function, will be proposed to complement or replace the power function approach. This exponential type function has nearly all the advantages of the power function, but not some of its disadvantages. Hence, it should become a desirable alternative, and may, in addition, generate research interest to develop additional tools and techniques in this area.

To obtain a proper perspective of, and relationship between, the two algebraic functions, the power function will be reviewed first, to be followed by the exponential type function.

POWER FUNCTION FORMULATION

Learning curve literature commonly starts out with the average cost function or more specifically, the cumulative average cost function; this chapter shall take a different approach for review purposes, and start out with the total cost function.

$$TC(x) = \frac{a}{1-b} \; x^{1-b} \; , \qquad (13\text{-}1)$$

where TC(x) is total cost or time incurred up to and including the x th unit,[9] x is the cumulative number of units
produced since the start of the process or operation, and
a and b are empirically-based parameters. The total cost
can be expressed in actual cost figures, or it can be expressed as an index, with the cost of the first unit being,
say, equal to one.

Average Cost

    If total cost is known, then average cost, or cumulative
average cost, can be obtained as follows;[10]

$$\frac{a}{1-b} \; x^{-b} \; , \qquad (13\text{-}2)$$

---

[9]Only "cost" will be used from here on instead of
"cost or time."

[10]The power function used has a negative exponent for
marginal and average cost with the result that the cost index declines with increasing productivity. If a positive exponent were used, as some authors do, then the index would
rise with increasing productivity, and would be called a
productivity or efficiency index.

314

where AC(x) is the average cost of the first x units produced in actual figures or measured by the same index used for total cost.

The latter formula is the one commonly known as the "learning curve" function, although it is directly related to the total cost function, and the marginal or unit cost function to be discussed next.

Marginal Cost

If the total cost function is differentiated the marginal or unit cost function results,

$$MC(x) = \frac{d[TC(x)]}{dx} = ax^{-b} \quad, \qquad (13-3)$$

where MC(x) is marginal or unit cost per unit for the xth unit[11] in actual cost or in total cost index terms.

---

[11]The marginal cost calculated by the mathematical formula is obtained from the slope of the total cost function. Since this total cost function is not linear (does not have a uniform slope) the marginal cost values obtained by the marginal cost function tend to be undervalued, which is especially significant at the start of the process. To compensate for this undervaluation, it is suggested to use the approximation

$$\overline{MC}(x) = \frac{MC(x-1)+MC(x)}{2} \quad,$$

or

$$\overline{MC}(x) = TC(x) - TC(x-1), \quad x = 2, 3, \ldots$$

For x=1, marginal cost, total cost, and average cost are the same.

The parameter b, in the equations (13-1), (13-2) and
(13-3) is constrained as follows,

$$0 \leq b < 1$$

If it were not, unrealistic values will be obtained. If
b=0, average cost and marginal cost are the same, and total
cost will increase at a constant rate. If b=1, then average
cost is undefined. There is no constraint on the value of
the parameter a; its value depends on the cost index used.

Disadvantages of Power Function

The disadvantage of the above power function formulation
can now be illustrated. Analyzing the marginal cost and
average cost functions, we notice that with increasing x,
both marginal and average cost will continue to decrease.
This may be a desirable condition for a cost-conscious firm,
but it is certainly not realistic, and especially not for
marginal cost. In actual practice average cost will continue
to decrease after marginal cost becomes a constant but at
a very slow rate. In the next section with an alternative algebraic formulation marginal cost will become a constant
after a certain number of units have been produced.

EXPONENTIAL FUNCTION FORMULATION

The proposed alternative to the power function formulation is an exponential function; it has been derived from the theory of difference equations. If the marginal cost of the xth unit, MC(x) is defined as y_{x-1}, then the theorem in the next paragraph can be applied.[12]

The linear first-order difference equation,

$$y_{x+1} = ay_x + b \qquad x = 0, 1, 2, \ldots \qquad (13\text{-}4)$$

(with y_o given and a and b empirically-based parameters) has the unique solution,

$$y_x = \alpha a^x + \beta \qquad 0 < a < 1 \qquad (13\text{-}5)$$

and $\{y_x\}$ converges with limiting value y^*. The parameters α and β are empirically-based, and

$$y^* = \frac{b}{1-a} \qquad (13\text{-}6)$$

---

[12]This theorem is a classical one and can be found in most text books on difference equations. The author suggests: Samuel Goldberg, <u>Introduction to Difference Equations</u>, New York: John Wiley and Sons, Inc. Science Editions, 1961.

The marginal cost of the first unit, $MC(1) = y_o$ is assumed known, and will be set equal to the index value one. Hence, from equation (13-5) note that,

$$y_o = \alpha + \beta = 1.00 \tag{13-7}$$

Furthermore, from equation (13-5) also note the convergence of $\{y_x\}$. With large x, a^x will approach zero, and

$$y_x = \beta = y^* \tag{13-8}$$

To estimate the parameters of equation (13-5) specify the value of y at convergence, and have one observation of y_x, or estimate one observation of y_x, for

$$y^* < y_x < y_o \ .$$

In the appendix the parameters for equation (13-5) are derived.

Total Cost and Average Cost

The total cost function can now be derived from the marginal cost function,

$$MC(x) = \alpha a^{x-1} + \beta$$

by integrating across x as follows,

$$TC(x) = \int (\alpha a^{x-1} + \beta)\, dx + c$$

$$= \frac{\alpha a^{x-1}}{\ln a} + \beta x + c, \qquad (13-9)$$

where MC(x) is the marginal cost per unit index for the xth unit, TC(x) is the total cost index up to the xth unit in terms of the marginal cost per unit index,[13] α, β, and a are empirically-based parameters, and c is a constant to be derived after the parameters have been determined.[14]

Average cost or cumulative average cost can now be derived; it is

$$AC(x) = \frac{TC(x)}{x} = \frac{\alpha a^{x-1}}{x \ln a} + \beta + \frac{c}{x}$$

where AC(x) is the average cost of the first x units produced in marginal cost index terms.

---

[13] The total cost values obtained by equation (13-9) and the resultant average cost values tend to be overstated somewhat due to the fact that a continuous function is used. Therefore, it is suggested, when x is small and when accurate total cost values are required, to use the sum of the marginal costs instead of equation (13-9).

[14] The constant c can be derived by letting MC(1)=TC(1)= AC(1), which provides: $c = \alpha - \alpha/\ln a$.

Conversion to Start Up Function

The exponential learning function can be easily converted to a start-up or productivity index by substituting
$1-a^{x-1}$ in place of a^{x-1}. If P(x) is the productivity function,
then

$$P(x) = \alpha(1-a^{x-1}) + \beta$$

Maximum productivity is reached at the index level 1 (one),
and the productivity index for the first unit is β.

Application Problems

The exponential functions discussed in this section are
not any more difficult to apply than the common and well-
known power functions. Although the exponential functions
have twice as many parameters as the power functions, these
parameters are very easy to calculate. For new processes
or new operations, especially those with large amounts of
uncertainty, the parameter estimation problem presents the
same difficulty with the application of either function.
However, the exponential functions have the rational feature of levelling out, to a constant value, for marginal
cost. The power function continues to decrease as output

increases for both the marginal cost and average cost cases.

REVIEW OF LEARNING FUNCTIONS

Two of the more recently-proposed learning functions have the same levelling-out to a constant value feature as the proposed exponential function. DeJong[15] proposes a modified version of the power function,

$$MC = a [\beta + (1-\beta)x^{-b}]$$

where a and b are parameters, analogous to the power function parameters a and b, and β is the minimum level which marginal cost will approach with large x. β is thus analogous to the β in the proposed exponential function.

One of the drawbacks of the DeJong function is the difficulty in estimating the parameters a and b. β is assumed known. In the comparison of the various learning functions in the next section, the parameter a will be set equal to 1 (one).

Explanatory Function

Levy's learning function[16] presents a still **more** difficult parameter estimating problem. However, the parameter

[15] J.R. DeJong, op. cit., pp. 51-59.

[16] F.K. Levy, op. cit., pp. 136-54.

estimating problems cannot be used as a criticism of Levy's
function, since the function is intended to explain how
firms adapt to new processes. Levy's function has the form,

$$MC = [1/\beta - (1/\beta - x^b/a) e^{-cx}]^{-1}$$

where a and b are the parameters that are analogous to the
power function parameters, c is the third parameter to be
estimated, and β is analogous to β in the proposed exponential
function, and is assumed known. In order to compare the
Levy function with the other functions, the parameters a
and b will be set equal to the power function parameters a
and b. The function of the parameter c is then to flatten
the curve for large values of x.

Carr and Guibert Functions

Two other, although older, learning functions are the
ones proposed by Carr[17] and Guibert[18]. Carr argued that
the cumulative average cost curve is best represented by an
S-type curve. The initial concave portion of the curve is
explained by the common practice, in most production operations, of adding one or several men at a time during the
initial start-up of a production program. Hence, only rela-

---

[17] G.W. Carr, op. cit., pp. 76-77.

[18] P. Guibert, op. cit.

tively small cost savings are obtained during the initial
start-up with the larger savings coming at a later production
stage. Since Carr's learning function applies to a special
case we shall not consider it in our analysis.

Guibert introduced the rate of production as a variable
affecting unit labor cost. He also viewed the cost curve
as having a horizontal asymptote that is approached after
a large number of units have been produced. However, the
rate of production is also viewed as a determinant of where
the decline in unit cost becomes negligible. The assumption,
which also underlies our exponential model, that unit cost
will eventually follow a horizontal asymptote is general.
However, tying it to the rate of production requires a model
that becomes restrictive. In addition, Guibert's model
centers around the notion that total number of units produced prior to the period in which peak output is reached
is approximately equal to the number of units in process
during peak production. Because of all the above restrictions
and assumptions, the model Guibert has developed is complicated and not very general. It is, therefore, not comparable
with the models being discussed in this chapter.

Economists and Production Functions

The economists have become interested in learning functions because of the effect learning has on production cost

functions. It is of interest to point out some of the
economists' contributions here as they relate to points
usually overlooked by contributors to the traditional learning curve literature. Alchian[19] agrees with the general proposition that cost of future output declines as total output produced increases. However, he restricts the proposition
to apply only to the increase in knowledge and not to a change
in technique. The above distinction is usually not made in
the industrial and production management literature.

Hirschleifer[20] responds to some degree to Guibert's
model by stating that rate of output will affect cost, but
in a different way. He postulates that with volume held
constant, marginal cost is a rising function of output.
However, with rate of output held constant, marginal cost
is a falling function of volume of output. On the latter
postulate all of the learning curve literature is based.

Orr[21] clarified not only Alchian's and Hirschleifer's
contributions, but also illuminated what has been presented

---

[19]Armen Alchian, "Costs and Outputs," The Allocation of
Economic Resources, Stanford, California: Stanford University
Press, 1959.

[20]Jack Hirschleifer, "The Firm's Cost Function: A Successful Reconstruction," The Journal of Business, Vol. 35,
July 1965, pp. 235-55.

[21]Daniel Orr, "Costs and Outputs: An Appraisal of
Dynamic Aspects," The Journal of Business, Vol. 37, January
1964, pp. 51-60.

by the traditional learning curve contributors. He starts
out by questioning whether any and all production activities
can be subsumed in Alchian's model because many firms can
modify operations through planning activities which may
affect cost considerably. He concludes by stating that
Alchian's theory and analysis is of little value in the
general case but may hold in industries such as aircraft,
job shop, etc., i.e., in all those industries where firms
produce on order with specified delivery dates. The latter
conclusion also applies directly to, and clarifies, all the
traditional learning curve literature. Learning curves have
not a universal application.

COMPARISON OF EXPONENTIAL FUNCTION WITH OTHER FUNCTIONS

The proposed exponential function will be compared with
the traditional power function and the two modified power
function formulations discussed in the previous section.
The data used for the comparison consists of eight sets of
production data collected by Levy for two offset pressmen
on a new offset press. The data is reproduced in Table 13-1,
in modified format. Levy's data showed actual productivity
for each period, whereas marginal cost data is used for our models.[22]

---

[22]In order to use the Levy data the average cost for
each period is considered to be the marginal cost at the
cumulative output level for that period.

Table 13-1

Production and Cost Data for Two Pressmen on New Press

Pressmen	Period	Cum.Output (Thousands)	Marginal or Unit Cost Index During Each Period			
			First Shift		Second Shift	
			With make ready	Without make ready	With make ready	Without make ready
1	1	101	1.000	1.000	1.000	1.000
	2	204	.951	.982	.987	.925
	3	286	.924	.956	.981	.861
	4	376	.928	.929	.959	.815
	5	460	.898	.903	.962	.812
	6	582	.871	.885	.946	.817
	7	682	.874	.853	.949	.810
	8	767	.835	.842	.910	.806
	9	852	.781	.845	.909	.792
	10	956	.776	.839	.904	.772
	11	1046	.774	.826	.897	.776
	12	1120	.758	.821	.904	.769
2	1	112	1.000	1.000	1.000	1.000
	2	202	.980	.971	.998	.973
	3	296	.939	.991	.956	.935
	4	390	.906	.908	.912	.862
	5	504	.909	.895	.914	.878
	6	590	.897	.901	.888	.848
	7	676	.893	.886	.885	.848
	8	786	.886	.859	.873	.817
	9	878	.886	.845	.879	.806
	10	994	.862	.820	.851	.809
	11	1090	.860	.819	.845	.800
	12	1164	.862	.801	.831	.801

The parameters for all four models were estimated from the
marginal cost and cumulative output data for the first and
seventh period. This method of estimating was used because
it more nearly approximates actual conditions than parameter
estimates based on all observations and derived by the least
squares method. Various values of β were used to ensure
that an unbiased comparison would result. A typical set of
parameters is shown in Table 13-2.

To obtain a meaningful comparison of the four functions,
the sum of squares of deviations (deviations of the actual
data from the values based on the algebraic function) for
each set of data were calculated. The means and standard
deviations of the sums of squares were also calculated and
are listed in Table 13-3 for three different sets of β
values.

CONCLUSIONS

The derived data supplies some interesting and valuable
information. In Table 13-2 we note that the parameter a
for the exponential model varies by only a small amount for
the eight sets of data. This indicates that the parameter
estimation problem may be simplified considerably for similar
operations. In addition, the more sensitive parameter b is
inversely related to the parameter a in the eight sets of data.
This property may be quite valuable if it holds true in all

Table 13-2

Typical Set of Empirically-Based Parameters

Series	β*	Power Model		Exponential Model	
		a	b	a	b
1	.734	1.385	.071	.9991	.0007
2	.803	1.468	.083	.9980	.0016
3	.894	1.135	.027	.9990	.0009
4	.746	1.664	.110	.9980	.0015
5	.848	1.346	.063	.9982	.0015
6	.781	1.374	.067	.9989	.0009
7	.814	1.378	.068	.9986	.0012
8	.781	1.542	.092	.9983	.0014

Series	DeJong Model		Levy Model**		
	a	b	a	b	c
1	1.000	.098	1.385	.071	0
2	1.000	.210	1.468	.083	0
3	1.000	.101	1.135	.027	0
4	1.000	.211	1.664	.110	0
5	1.000	.187	1.346	.063	0
6	1.000	.113	1.374	.067	0
7	1.000	.148	1.378	.068	0
8	1.000	.182	1.542	.092	0

* β is based on a typical γ-value of 0.10, as expressed in the equations: β=γ [cost(period 1)-cost(period 12)].

** The value of parameter c was negligible and caused the Levy model to become a replica of the traditional power model.

Table 13-3

Measures of Actual Data Deviating from Algebraic Functions

Function	γ*	Sum of Squares of Deviations	
		Mean	Standard Deviation
Power	−	.179	.324
	−	.179	.324
	−	.179	.324
Exponential	.02	.039	.026
	.10	.039	.026
	.20	.039	.026
DeJong	.02	.065	.032
	.10	.064	.031
	.20	.063	.031
Levy	.02	.179	.324
	.10	.179	.324
	.20	.179	.324

* The constant γ is used to vary β in the equation:

$$\beta = \gamma \,[\text{cost(period 1)} - \text{cost(period 12)}]$$

cases. Table 13-3 indicates that the proposed exponential function provides a considerably better and consistently better fit (smaller mean and smaller standard deviation) than the three other functions. Only the DeJong function approaches the proposed exponential function. Another important and valuable feature of the proposed exponential function is the fact that varying the value of β has no observable effect on the fit of the algebraic function to the data. The latter feature is important, since β will have to be estimated, and the basis for the estimate will commonly be quite vague.

EXERCISES

1.   The Eagle Aircraft corporation is a builder of
aircraft for commercial and military purposes.   From
historical data it has found that the cost of the sixth
aircraft completed is about 1/3 of the cost of the first
aircraft.   It has also been determined that cost per
aircraft built does not level out until the twentieth
aircraft is built.   Find the parameters of the classical
learning (power) curve for Eagle Aircraft.

2.   Based on the information in problem 1 find also the
parameters for the exponential learning curve.

3.   Compare the projected cost of the first thirty aircraft on the basis of the two models.   The cost of the
first aircraft is $3,900,000.   What is the cost of the
twentieth aircraft?

4.   Yoda Electronics has found by experience that production cost of electronic household entertainment products such as stereo sets, television sets and so forth
decreases over time on the following basis.   If cost
per unit produced for the first one hundred units has
an index of 100 then cost per unit for the tenth one
hundred units has an index of 60 and cost standardizes

to an index of about 25 after 5000 units have been produced. Find the parameters for both models on the
basis of the above information.

5. Compare the cost projections for the two models for
the first 15,000 units in problem 4.

6. Show that when the effect of learning is neglected,
the break even point is higher.

7. A firm finds that it costs $300 to produce the first
165 items. If it takes an additional $180 to produce
another 165 items, how much will it cost to produce 1320
items?

8. The following algebraic functions for total cost,
average cost and marginal cost appear to be excellent
substitutes for the learning functions discussed in
this chapter. Discuss their relative merits and disadvantages.

$$TC = c - a \exp(-bx) / b$$
$$AC = c/x - a \exp(-bx) / (bx)$$
$$MC = a \exp(-bx)$$

9. If you were a contract administrator for an advanced
systems firm or for a federal agency, how important
is the concept of learning curves to you in deciding
the actual costs to contractors of final bid contracts?

10. Space Dynamics specializes in aerospace assemblies. It has found over time that assembly labor is reduced considerably as more units of a given model are produced. In fact it has found that each time production doubles average labor cost per assembly is reduced to 80 percent of what it was previously. To estimate its total assembly labor cost and also its output rate Space Dynamics has applied the learning function known as the power function. With an 80 percent average labor cost reduction rate the parameter b in the mathematical learning function

$$AC(x) = \frac{a}{1-b} x^{-b}$$

can be found from the formula

$$b = -\log .80/\log 2$$

To determine the parameter a Space Dynamics depended on its methods engineering department to determine how much time it took to assemble a unit after the assembly workers had become familiar with the assembly and fully trained. The time and cost at that point was assumed to have reached a constant value for each additional unit produced. Hence to determine a the marginal cost formula,

$$MC(x) = a x^{-b}$$

was used. The formula thus arrived at had performed an excellent job for Space Dynamics so far. Total cost could be determined at any point during the contract period from the formula,

$$TC(x) = \frac{a}{1-b} x^{1-b}$$

and cost estimates were amazingly accurate. However, management had become concerned because beyond the point where marginal cost becomes a constant the formula had been applied in error in the past and estimates had been used which grossly underestimated cost. To avoid this management was searching for a formula which would take into account the marginal cost levelling out feature and also produce accurate estimates at other production levels.

Make a presentation for mangement of Space Dynamics which will compare the power curve and the exponential curve for production of 800 transponders. The assembly cost of the 240th transponder is estimated at $80 per unit and an 80 percent learning curve has been found to be applicable in the past.

11. The previous problem suggests the use of the power curve to determine the parameters of the exponential curve. You are asked to explain this procedure in a set of procedures for use by cost analysts, purchasing clerks, etc. Use an illustration to explain the procedure. For instance use a production level at which marginal cost stabilizes to a constant value; then determine the power curve and from the power curve find the parameters for the exponential curve.

12. Explain why you would want your purchasing agents to be familiar with the learning curve applications and limitations if you were the general manager of an aerospace firm.

APPENDIX

From equation (13-4), for $x = 1$,

$$y_1 = ay_0 + b,$$

and for $x = 2$, it follows that,

$$y_2 = ay_1 + b = a^2y_0 + b(a+1),$$

and for any x,

$$y_x = a^x y_0 + b \sum_{i=0}^{x-1} a^i$$

But $y_0 = 1.00$, and the geometric progression can be summed, so that we obtain,

$$y_x = a^x + b\left(\frac{a^x - 1}{a-1}\right) \tag{A13-1}$$

From equations (13-6) and (13-8) we can derive,

$$b = \beta(1-a), \tag{A13-2}$$

and from equations (A13-1) and (A13-2) we find that

$$a^x = \frac{y_x - \beta}{1-\beta}, \tag{A13-3}$$

and,

$$a = \text{antilog}\left[\frac{\log(y_x - \beta) - \log(1-\beta)}{x}\right]$$

Suppose that y_x converges at $y^* = \beta = 0.20$. Then, it follows from equation (13-7) that $\alpha = 0.80$. Also suppose that $x=60$, $y_x = .60$. Then from equation (A13-3) we find that a = 0.9935.

To determine the value of y_x at a given value of x we use the total cost incurred up to the (x+1)th unit processed or produced and subtract from it the total cost incurred up to the xth unit.

CHAPTER 14.--FORECASTING DEMAND ON THE BASIS OF EXOGENOUS

        FACTORS

    This chapter and the next one are concerned with fore-
casting demand for an individual product or for a firm's
aggregate demand. There are basically two types of forecast problems. On the one hand forecasts may be based on
historical demand data of the item itself. This approach
will be explored in the next chapter. On the other hand
demand forecasts may not at all be related to the historical
demand and must therefore be based on exogenous factors.

        EXAMPLES OF EXOGENOUS FACTORS

    The use of exogenous factors to forecast sales is so
common that its use is essentially never questioned. This
is not surprising because the exogenous factors that are
used are almost always causally related to sales of the items.
In the few cases where the relationships cannot be causally
explained the use of unrelated exogenous factors should be
avoided.

    Exogenous factors that are commonly used are the pre-
dicted state of the economy or the predicted demand of a related item. For instance, the demand for new bathroom fix-
tures will be dependent on the demand for new houses, similarly

the demand for automotive windshield wiper motors will be a function of the demand for new automobiles.

It should be clear to the reader that forecasting demand on the basis of some exogenous factor increases the magnitude of the potential forecast error. Consider the problem of predicting demand for windshield wiper motors, which in turn depend on the demand for automobiles. To be specific, the common approach used in such a problem is to determine the relationship, on the basis of historical data, between the demands for windshield wiper motors and automobiles. In the rather simple example one may find that the relationship is just a simple one-to-one ratio. In other cases, it may be a more complex relationship. If it is a one-to-one ratio then it is rather simple to determine the expected demand for windshield wiper motors. It is exactly the same as the expected demand for automobiles during the same time interval. However, the accuracy of the automobile demand forecast only depends on the accuracy of the method used to forecast automobile demand. The accuracy of windshield wiper motor demand, on the other hand, depends on both the accuracy of the estimated ratio of windshield wiper motors to automobiles and on the accuracy of the automobile demand forecast.

From the above example it should be clear that in practice, predictions are often considerably in error, both because it is difficult to obtain accurate estimates of the relationship between the two variables, windshield wiper motor demand and automobile demand, and because it is often extremely difficult to predict automobile demand. It is not unusual to add another variable to the above problem. Suppose there is an established relationship between automobile demand and net disposable income of the economy. In that case windshield wiper motor demand is related to automobile demand and automobile demand is related to net disposable income. An obvious shortcut in such a case is to attempt to find a direct relationship between windshield wiper motor demand and net disposable income.

In the above discussion no attempt was made to introduce probabilistic considerations, except to state that demand will be a random variable. The problem is that, in the above example, not only windshield wiper demand is a random variable, but also the relationship between automobile demand and windshield wiper demand is a random variable even if automobile demand were deterministic instead of stochastic. Therefore, the usual procedure in such a case is to assume that the demand random variable is distributed according to some theoretical distribution such as Poisson, Normal, Negative Binomial, and so forth. Before discussing more sophisticated methods for demand forecasting let us first take a brief look at the more mundane approach of subjective opinion surveys of management.

OPINION POLLING[1]

Opinion polling is probably the most well-known statistical technique, especially during election years. The opinion poll or sample survey technique of forecasting is a subjective method of prediction, amounting largely to an averaging of attitudes and expectations. The underlying assumption is that certain attitudes affecting economic decisions can be defined and measured well enough in advance so that predictions can be made of changing business or sales trends. The results are arrived at by asking people who are directly involved about their expectations as to future economic events or states and the results of these states of the economy on the expected demand. Various forms or types of surveys are employed both in economic and in sales forecasting.

Management Opinion Poll

There are two basic ways in which opinion polling can be employed to forecast demand, directly or indirectly or both. With the direct way the views and opinions of management regarding future demand are collected and then used to

---

[1]See also: M.H. Spencer, Managerial Economics, Homewood, Illinois: Richard D. Irwin, Inc. 1968, pp. 59-68.

arrive at a forecast. With the indirect way the views of management, regarding future economic conditions in the industry or economic sector which determine sales, are collected. The resulting economic forecast is then used to arrive at a sales forecast.

The basic assumption of management polling is that there is safety in numbers. That is the combined or average forecast of the group is better than the forecast of any single member. Since each group member's forecast is usually recorded for future comparison against actual sales, this method has a reasonably high motivation for making the best subjective forecasts possible. To prevent bias, a divergence of opinions is usually obtained by including various divisional management groups such as the firm's sales, production, finance, purchasing, and administrative divisions.

The above approach can be quite successful if the subjective forecasts are based on objective information derived from analyzing market reports, historical sales data, forecasts of the economy, and so forth. Without a careful basis for subjective judgment, the subjective forecasts can easily become a guessing game which result in an inaccurate and useless forecast of sales.

Sales Force Forecast Opinion Poll

Another source of subjective sales forecasts is a firm's

sales force. The sales force is closest to the market and one would expect that the specialized knowledge of those nearest to the market is an invaluable asset in arriving at a good subjective forecast. Especially, if a structured approach is developed, whereby each salesman is required to contact all his current customers to obtain their estimates of their future needs, one can be quite confident about the individual demand forecasts. Individual branch sales and region sales managers can then further analyze the sales projections to ensure that a reasonable degree of uniformity is maintained. One possible drawback to this approach is that salesmen may be unaware of the structural changes which are taking place in the industries or market segments they are servicing. Hence, sales projections may be based on historical rather than future states of the market. The difference between the sales force polling and management polling is that all sales force sales projections are accumulated and the various parts make one aggregate forecast. However, the aggregate forecast is essentially only one observation if the sales polling is viewed as an experiment. The management polling on the other hand consists of many observations or estimates. This large number of estimates is then averaged by a simple or weighted averaging technique to arrive at an average sales forecast.

Customer Opinion Poll

One other polling technique which may be helpful to arrive at a forecast is the customer poll. It may consist of a direct poll to all customers or of a sample poll to a randomly selected subset of all customers or potential customers. The results of customer polls may, however, be highly biased and very dependent on certain events to take place. For instance, not every one knows for sure on January 1 whether he will purchase an automobile or not within the next twelve months. In addition the survey data so collected when aggregated consist of only one sales projection.

FORECASTS BASED ON EXOGENOUS FACTORS

This section is concerned with forecasts based on economic indicators. It has been found in many cases that improved forecasts can be obtained on the basis of indicators other than past sales. For instance it may be hypothesized that there is a relation between a building product item sales and general business activity indicators. The analysis can usually be limited to indicators which are available monthly, or at least quarterly on a seasonally adjusted basis. Five possible indicators are definite candidates for a building product item. These are:

    a.   Gross National Product (GNP) in constant dollars.

    b.   Disposable Personal Income deflated by the implicit deflator of the consumption component of GNP.

    c.   New Construction component of GNP in constant dollars.

    d.   Gross Private Investment component of GNP in constant dollars minus changes in inventories.

    e.   Federal Reserve Board Index of Industrial Production.

One observation immediately suggests it may be more desirable to use regional economic indicators, especially if the bulk of sales comes from one or several specific regions. Unfortunately, regional economic indicators are not readily available nor are they very usable. However, the general level of economic activity in many regions tends to be closely related to national economic activity, and therefore national indicators, which are easier to obtain, often serve nearly as well.

The sales forecaster's task is to determine which of five economic indicators he should choose. This can usually best be accomplished by comparison with historical data to determine the degree of correlation between these indicators and the actual sales level. The analysis may still be further refined by evaluating quarterly or half yearly indicators.

Of course, looking backward to compare performance of various indicators is not the same as making forecasts on the basis of the indicators. Suppose it is found that an economic indicator is closely correlated with the sales of

a product. The problem faced then is how to obtain accurate
forecasts of the indicator for the year for which the sales
forecast is to be made. Fortunately, reasonably reliable
forecasts can usually be obtained from the various univer-
sities, banking and life insurance institutes who prepare
annual and quarterly forecasts. Forecasts are usually especially
accurate for the aggregate economic indicators such as GNP
and Disposable Income. Forecasts for some of the smaller
segments sometimes leave something to be desired. Therefore,
in choosing an economic indicator as independent variable
for the sales forecast, it is wise to remember that the percentage error between the various economic indicators varies
considerably.

FORECASTING DEMAND INTERVALS BY ANALYTIC METHODS[2]

How to make a sales forecast on the basis of an independent variable will be discussed in this section. This
independent variable may be one of the economic indicators
discussed in the previous section or some other indicator.
It will be assumed that the dependent variable is normally
distributed and that the historical observations of the independent variable are uncorrelated. Hence, the model is

---

[2]See also: A.M. Mood and F.A. Graybill, Introduction
to the Theory of Statistics, New York: McGraw-Hill Book
Company, Inc. 1963, pp. 328-38.

a simple linear model.

Let y be the demand for the commodity and x the magnitude of the indicator at the same period. A demand forecast must be made for a future period, commonly the next period. Observations are collected on the pairs (x_i, y_i) over a recent historical period up to the latest period for which actual information for the variables is available. The demand variable y_i is the demand for period i, and x_i is the magnitude of the indicator for period i.

For a simple linear model in which the random variables are normally distributed the likelihood estimators $\hat{\alpha}, \hat{\beta}$ and $\hat{\sigma}^2$ can be used as estimators for the parameters α, β and σ^2. These are obtained from the following formulas,

$$\hat{\beta} = \Sigma(x_i - \bar{x})(y_i - \bar{y}) \; / \; \Sigma(x_i - \bar{x})^2$$

$$\hat{\alpha} = \Sigma y_i/n - \hat{\beta} \; \Sigma x_i/n$$

$$\hat{\sigma}^2 = \Sigma(y_i - \hat{\alpha} - \hat{\beta}x_i)^2/n$$

From the above estimators a forecast can be made what demand will be in period k by the formula,

$$\hat{y}_k = \hat{\alpha} + \hat{\beta} \; x_k \; .$$

This prediction is not very useful unless there is some knowledge about its possible error. If the magnitude of the error is known then a prediction interval can be determined which essentially indicates that sales will range between a level $\hat{y}_k \pm K$ with an assurance of a certain specified percentage between 0 and 100.

The variable y_k is a random variable with a normal distribution as assumed. It has a mean $\alpha + \beta x_k$ and variance σ^2. There are two sources of error in the forecasted value $\hat{y}_k = \hat{\alpha} + \hat{\beta} x_k$. First, $\hat{\alpha} + \hat{\beta} x_k$ is only an estimate of the mean of the random variable y_k, and the actual value of y_k may deviate from its mean. Secondly, the estimated mean is subject to the random sampling errors inherent in $\hat{\alpha}$ and $\hat{\beta}$. If α, β and σ^2 were exactly known then a 95% assurance interval for the forecasted demand will be,

$$\alpha + \beta x_k - 1.96\sigma \leq y_k \leq \alpha + \beta x_k + 1.96\sigma$$

However, only x_k is known and the parameters must all be estimated. Hence, an interval estimate must be set up for them also. This can be accomplished by letting,

$$v = y_k - \hat{\alpha} - \hat{\beta} x_k$$

The variable v is also normally distributed since it is a linear function of three normally distributed variables. The variance of v is therefore,

$$\sigma_v^2 = E(v^2)$$

$$= E(y_k - \hat{\alpha} - \hat{\beta} x_k)^2$$

$$= \sigma^2 + \sigma_{\hat{\alpha}}^2 + x_k^2 \, \sigma_{\hat{\beta}}^2 + 2x_k \, E\,[(\hat{\alpha}-\alpha)(\hat{\beta}-\beta)] \qquad (14\text{-}1)$$

Since,

$$\sigma_{\hat{\alpha}}^2 = \sigma^2 \, \Sigma x_i^2 \, / \, [n\Sigma(x_i - \bar{x})^2]$$

$$\sigma_{\hat{\beta}}^2 = \sigma^2 / \Sigma (x_i - \bar{x})^2$$

$$E[(\hat{\alpha}-\alpha)(\hat{\beta}-\beta)] = -\sigma^2 \bar{x}/\Sigma(x_i - \bar{x})^2$$

formula (14-1) can be simplified to,

$$\sigma_v^2 = \sigma^2 \left[(n+1)/n + (x_k - \bar{x})^2 / \Sigma(x_i - \bar{x})^2 \right]$$

A 95 percent assurance interval for v is thus,

$$-1.96\sigma_v \leq v \leq 1.96\sigma_v \ ,$$

but this still involves the unknown parameter σ_v. However, v/σ_v is normally distributed with zero mean and variance one, and is independently distributed of $n\hat{\sigma}^2/\sigma^2$. As a result σ can be eliminated by using the t distribution as follows,

$$t = (v/\sigma_v) \, / \, \{n\hat{\sigma}^2 \, / \, [(n-2)\sigma^2]\}^{\frac{1}{2}}$$

with n-2 degrees of freedom.

The $100\text{-}100\varepsilon$ assurance interval for t is therefore,

$$-t_{\epsilon/2} \le t \le t_{\epsilon/2}$$

which may be converted to a 100-100ε assurance interval for $\hat{y}_k$ as follows,

$$\hat{\alpha} + \hat{\beta}x_k - z \le y_k \le \hat{\alpha} + \hat{\beta}x_k + z$$

where

$$z = t_{\epsilon/2} \hat{\sigma} \{ n/(n-2) [(n+1)/n + (x_k - \bar{x})^2 / \Sigma (x_i - \bar{x})^2] \}^{\frac{1}{2}}$$

Before presenting an example of the above technique, the reader should be aware that the above method only applies to linear relationships between sales and a selected indicator. If there appears to be a non-linear relationship other methods must be used to forecast demand. Non-linearity will usually be indicated by very wide forecast intervals for high assurance levels. Such forecasts are not very valuable and other forecast methods should then be applied.

AN EXAMPLE OF INTERVAL ESTIMATING

Suppose data are available on the sales of bootstraps for the Oostra Bootstraps Company as shown in Table 14-1. The last column indicates an economic indicator which is causally related to bootstrap sales, provides a high correlation with bootstrap sales, and also provides a reasonably good linear relationship. The linear model discussed in the previous section will be applied. The y variable is assumed to be normally distributed and a forecast interval for the year 1969 of bootstrap sales is desired.

Table 14-1

Historical Sales and Economic Indicator Level

Period	Historical Sales (y_i)	Economic Indicator (x_i)
1954	91	15
1955	102	16
1956	99	18
1957	108	17
1958	106	18
1959	109	19
1960	111	19
1961	114	20
1962	110	22
1963	112	21
1964	113	22
1965	116	23
1966	119	24
1967	123	25
1968	124 (estimated)	26 (estimated)
1969		27 (forecast)

Applying the various formulas in the previous section the forecast for 1969 bootstrap sales is found to be,

$$\hat{y}_{69} = 127 \text{ (rounded; actual value is 126.62)}$$

and the forecast interval for a 95% assurance level is

$$117.9 < y_{69} < 135.3.$$

SHORTCUT TO INTERVAL ESTIMATION

If sales forecasts of many products have to be made on a frequent basis the method discussed above may become somewhat unwieldy and costly. Therefore there is a need for a simpler, less costly and faster variance estimation technique.

One obvious shortcut is to assume that the variance is a constant over a fairly wide interval of sales and to use the historical variance applied to the sales forecast as a predictor for the sales forecast interval.

If sales do not vary much over time the assumption of constant variance is a satisfactory method. However, if sales do vary over time research by Holt et. al.[3] has indi-

---

[3]C.C. Holt, F. Modigliani, J.F. Muth, and H.A. Simon, Planning Production, Inventories and Work Force, Englewood Cliffs, N.J.: Prentice-Hall, Inc., 1960, pp. 274-75.

cated that the assumption that variance is constant is in-
valid. However, there is a statistical parameter that
appears to remain constant over time and that parameter is
the coefficient of variation.

To study the relationship between the mean and variance
of historical sales Holt et al calculated the means and
standard deviations of a class of products over eighteen
monthly periods. The means and standard deviations of each
product were then plotted and a coefficient of variation
of 0.8 was indicated.

The above coefficient of variation constant is of
course only applicable to a specific class of products, but
one could hypothesize that a class of products could be very
broad such as to include all household product, home building
and general construction products. Further research in
this area seems warranted, but in the meantime the use of a
constant coefficient of variation seems appropriate for forecasting sales intervals if cost and speed are important factors.

EXERCISES

1. A high degree of correlation has been found to exist
between sales of a specific brand of household detergent
and the price index for pork bellies on one of the mercantile
exchanges. Would you suggest management take advantage of

this or should a study of this relationship be continued?

2. Widget sales have been observed to be related to the new housing starts index. Historical sales of widgets in millions of units and the historical index of housing starts are shown below.

Year	Widget sales in millions	Housing starts index
1960	350	11.30
1961	315	10.26
1962	370	12.11
1963	365	12.23
1964	405	13.04
1965	335	12.90
1966	385	13.14
1967	425	14.05
1968	390	13.25
1969	365 (estimated)	11.96 (estimated)
1970		10.40 (forecast)

Calculate sales forecast interval for widgets for 1970.

3. The coefficient of variation for annual sales of garden tools has been found to be 0.75. A forecast for a special brand of shears has been made and amounts to 560,000 units per year. If sales of shears is normally distributed what

is the sales forecast interval for a 95% assurance level?

4. A sales forecaster hypothesized the following regression equation to explain past sales,

$$y = 5000 + 31x_1 + 12x_2 + 49x_3$$

where,

  y = yearly sales in units of $1,000

  x_1 = U.S. disposable personal income in billions of dollars,

  x_2 = U.S. population in millions of households, and

  x_3 = time in years away from 1960, 1960 = 0

This equation gives R^2 = .98. What key assumptions must have been made for this equation? Would you accept this equation as being reasonable?

5. Sales of electric power by Valley Power and Light has increased during the past 60 years as shown below

Year	Sales in million kilowatt hours
1910	100
1920	285
1930	580
1940	910
1950	1980
1960	4260
1970	7940

a. Find the average amount of increase per decade between 1910 and 1970.

b. Find the average percentage increase per decade between 1910 and 1970.

    c. Build a regression model to forecast 1980 power sales, use the log of power sales as the independent variable.

6. If you were the president of Valley Power and Light would you accept the prediction of power sales for 1980 based on the above model?

7. The relationship between average inventory levels, which are estimated at the end of each quarterly period, and annual sales of General Department Stores can be estimated from the data shown below.

	Millions of Dollars	
Year	Average Inventory	Annual Sales
1961	23.0	159
1962	22.4	158
1963	24.3	174
1964	25.9	179
1965	26.1	190
1966	25.8	192
1967	26.0	206
1968	28.7	210
1969	28.4	208
1970	28.9	215

Management of General Department Stores projects sales for the next three years to increase by eight percent each year and it would like to know what its capital needs are for the increased inventory needed to support the increased sales levels.

8. Discuss the forecasting problems of a job order company which only manufactures to orders. No inventory is kept. What kinds of forecasts must this company make? Is it feasible for this company to inventory any of its resources during slack periods?

9. Sales of candy and greeting cards are heavily influenced by special holidays such as Christmas and Easter. The date for Christmas is fixed but the date for Easter changes from year to year. In three consecutive years the dates for Easter were April 22, April 14 and March 13. How would you take into account this change while making monthly or quarterly forecasts?

10. Prior to an election a polling organization claims that its poll shows that a certain candidate will win. Another polling organization claims that another candidate will win. Following the election which was won by a third candidate the two polling organizations make a joint statement which claims that the results of the election are in agreement with the results of their polls although neither one of the two picked the winner. Explain how they could make such a statement.

CHAPTER 15.--FORECASTING DEMAND BY EXPONENTIAL SMOOTHING
                METHODS

        In the previous chapter demand forecasting was dis-
cussed for individual products or for the firm's aggregate
production on the basis of what is happening or is expected
to happen in the external environment. In this chapter
sales forecasts based on what has occurred in the past
will be discussed. The methods to be concentrated on in
this chapter are especially appropriate when forecasts
must be made cheaply, easily, routinely, and preferably
by electronic computer. The need for this type of forecasting technique is especially important when forecasts
must be made for large numbers of products such as occurs
in the replacement parts market and so forth.

                    MOVING AVERAGES

        The simplest routine forecasting method which uses
historical information is probably the moving average forecast. If there are no cyclical or seasonal variations and
no significant upward or downward trend in sales a moving
average forecast is just a simple or weighted average of
sales over the past n periods. The value of n must be
determined by the forecaster. If there is a definite upward or downward trend in sales it is not too difficult to
incorporate this into the moving average forecast. However, if there is a seasonal or cyclical variation it

becomes a bit more difficult to handle the moving average forecast. It is still possible to do so but it is questionable if the moving average forecasting technique satisfies the general specification that forecasts be made cheaply, easily and routinely. With seasonal or cyclical sales variations the sales history of a considerably longer period must be stored and therefore the computer memory storage cost would go up sharply. For instance, let us take a look at the example in Table 15-1. The basic concept for this method is borrowed from Holt, Modigliani, Muth and Simon.[1]

Based on historical monthly sales for the past two years, a fraction index of seasonals is developed which shows that eleven percent of total annual sales occurs in January, nine percent in February, and so forth. This fraction index or list of seasonals in the example is based on the average for two years, but can just as easily be based on more years. The forecast for the next period, January of year t is then the fraction for January times the total annual sales for year t-1. The moving average thus consists of twelve monthly periods. For the February forecast the sales for January of year t-1 (465 units) would be subtracted and the actual sales for January of

[1]C.C. Holt, F. Modigliani, J.F. Muth, and H.A. Simon, Planning Production, Inventories and Work Force, Englewood Cliffs, N.Y.: Prentice-Hall, Inc., 1960, p. 135.

year t would be added. The resultant twelve months total
would then be multiplied by 0.09 to obtain the forecast
for February of year t.

The last column of Table 15-1 shows a different
forecast. This forecast takes into account an annual
upward trend in sales. This upward trend is based on the
actual annual sales over the past five years shown in
Table 15-2. This five year trend shows an average annual
increase of eight percent per year. The no trend forecast
in the second last column is multiplied by 1.08 to obtain
the forecast with the trend included. To obtain the trend
forecast for February do not use the 1.08 multiplication
factor but 1.08- .08/12, or 1.0733. Similarly, for March
multiply the no trend forecast by 1.0667, and so forth.
At the end of year t recalculate the average annual sales
increase or trend factor by dropping year t-5 and adding
actual sales for year t.

The method discussed above is thus quite versatile
and can be used with or without cyclical or seasonal sales
variations, with or without trend effects, or with both
cyclical or seasonal sales variations and trend effects.
There is, however, one serious problem associated with
the above technique. That problem is that the moving
average method requires too much memory storage for a
computer application. It is fine for a manual application,

Table 15-1

Historical Sales for Forecast Purposes

Period	Year t-2	t-1	2-year totals	Seasonals Fraction of 2-year total	Forecast for year t (no trend)	Forecast for year t (trend)
1	360	465	825	.11	430	464
2	240	435	675	.09		
3	270	330	600	.08		
4	210	240	450	.06		
5	330	270	600	.08		
6	420	480	900	.12		
7	540	510	1050	.14		
8	210	165	375	.05		
9	240	210	450	.06		
10	120	180	300	.04		
11	240	210	450	.06		
12	405	420	825	.11		
	3585	3915	7500	1.00		

Table 15-2

Historical Annual Sales Trends

	Year t-5	t-4	t-3	t-2	t-1
Annual total sales	2880	3160	3275	3585	3915
Increase over previous year	——	280	115	310	330
Percentage increase	——	9.7	3.6	9.5	9.2

but will become quite costly if put into a computer for an application with a large number of items. Therefore, in the next section a method will be reviewed which will perform the same functions as the moving average method but also can be easily performed on a computer at low cost. The method is known under a variety of names such as exponential smoothing, exponential forecasting, adaptive forecasting, and exponentially-weighted moving averages.

EXPONENTIAL SMOOTHING

Exponential Smoothing by weighted moving averages is a technique for predicting sales of individual items by extrapolating sales time series. Its use is appropriate only in those cases where it is impractical to introduce information concerning the market, the industry, and the economy. Products that lend themselves to exponential forecasting are found in the market of replacement parts, hardware items, kitchen wares, and so forth. Hence, the area of application is the same as that discussed in the previous section.

The only computer memory required for an exponential forecasting system, once the system is operational, are for the values of the parameters of the particular forecasting formula and for a limited number of permanent forecast components. A forecast can then be obtained for each period, and this forecast is based on the historical time series, and on the actual sales of the most recent period. The latter must be supplied to the forecasting system each period.

The system described above was briefly mentioned by Magee[2] was discussed in more detail by Brown[3,4] and eloquently introduced by Holt et.al.[5]. Magee only discussed a basic model which could not handle seasonal or trend effects. Brown and Holt et.al. introduced the notion of seasonal and trend effects, but never completely presented all possible seasonal and trend effect combinations. However, it is imperative that the appropriate model be used, once it has been decided to use exponential forecasting.

At this juncture it may be appropriate to mention the notion of double and triple exponential smoothing proposed by some authors. These methods will not be discussed here because the use of cyclical or seasonal and trend modifiers accomplish the same purpose and in a way which

---

[2] J.F. Magee, Production Planning and Inventory Control, New York: McGraw-Hill, 1958, Chapter 6.

[3] R G. Brown, Statistical Forecasting for Inventory Control, New York: McGraw-Hill, 1959, Chapter 2.

[4] R.G. Brown, Smoothing, Forecasting and Prediction of Discrete Time Series, Englewood Cliffs, N.J.: Prentice-Hall, Inc., 1963, Chapter 7,8 and 9.

[5] Holt, et.al., op. cit., Chapter 14.

can be causally explained.

The Basic Exponential Model

For products which have no definite cyclical or seasonal pattern and no trend effects the basic exponential smoothing model can be used. The basic exponential model assumes that sales are made up of a weighted average of a random component which lasts a single period of time and a permanent component which lasts through all subsequent periods. The forecast of sales in each future period T is,

$$S_{t,T} = \bar{S}_t$$

where $T = 1,2,\ldots$; $\bar{S}_t$ is the estimate of the permanent component and $S_{t,T}$ is the forecast of sales to be realized in period t+T based on information available through period t. The forecast for the next period would thus be $S_{t,1}$.

The estimate of the permanent component changes each period, on the basis of additional sales information, by a fraction of the most recent forecast error. The permanent component for period t can then be expressed as,

$$\bar{S}_t = \bar{S}_{t-1} + w_e(S_t - \bar{S}_{t-1}), \qquad (15\text{-}1)$$

where S_t is actual sales in period t and w_e is less than or equal to one and greater than or equal to zero.

From formula (15-1) it follows that,

$$\bar{S}_{t-1} = \bar{S}_{t-2} + w_e(S_{t-1} - \bar{S}_{t-2})$$

which can also be expressed as,

$$\bar{S}_{t-1} = w_e S_{t-1} + (1-w_e) \bar{S}_{t-2} \qquad (15-2)$$

Substituting formula (15-2) into formula (15-1) results in,

$$\bar{S}_t = w_e S_{t-1} + (1-w_e)\bar{S}_{t-2} + w_e(S_t - \bar{S}_{t-1}), \qquad (15-3)$$

and substituting formula (15-2) into formula (15-3) results in

$$\bar{S}_t = w_e S_{t-1} + (1-w_e)\bar{S}_{t-2} + w_e S_t - w_e^2 S_{t-1} - w_e(1-w_e)\bar{S}_{t-2},$$

which can be simplified to,

$$\bar{S}_t = w_e S_t + w_e(1-w_e)S_{t-1} + (1-w_e)^2 \bar{S}_{t-2}. \qquad (15-4)$$

Note that formula (15-4) expresses the permanent component of sales in terms of a fraction of each of the two previous periods' sales and a fraction of the permanent component two periods ago. Continuing the above process the permanent component of sales $\bar{S}_t$ can be expressed explicitly in terms of all the past observations of sales as follows,

$$\bar{S}_t = w_e S_t + w_e(1-w_e)S_{t-1} + w_e(1-w_e)^2 S_{t-2} +$$

$$+ \ldots + w_e(1-w_e)^N S_{t-N} + (1-w_e)^{N+1} \bar{S}_{t-N-1}. \qquad (15-5)$$

Formula (15-5) can be rewritten in simpler form as,

$$\bar{S}_t = w_e \sum_{n=0}^{N} (1-w_e)^n S_{t-n} + (1-w_e)^{N+1} \bar{S}_{t-N-1} \qquad (15-6)$$

For large N the last term in formula (15-6) approaches zero and can be ignored. Hence, formula (15-6) becomes,

$$\bar{S}_t = w_e \sum_{n=0}^{N} (1-w_e)^n S_{t-n}. \qquad (15-7)$$

Note that formula (15-7) expresses the permanent component of sales in terms of a weighted average of past sales. The weights attached to past sales add up to unity and hence no systematic bias is introduced.

One problem in the above formula has not been resolved yet. That is the weight w_e must be determined independently of the formula. One way of determining w_e is by minimizing the square of the forecast error over an adequate set of historical data. This is a computational task of considerable magnitude. However, with the aid of a computer it becomes a relatively simple task. In practice, w_e is usually selected on the basis of how influential recent sales are in determining future sales. If recent sales heavily influence future sales then w_e should be relatively large; if on the other hand historical sales do not heavily influence future sales then w_e should be relatively small.

Cyclical or Seasonal Variations

If cyclical or seasonal changes occur in the sales level then these should be taken into account. Cyclical changes may be based on weekly, monthly or annual cycles and seasonal changes are usually the result of holiday periods or special activities or needs occasioned by a particular time of the year.

Cyclical or seasonal changes are recorded or stored in the cyclical or seasonal permanent component, F_t, which is updated each cycle of length L periods as follows,

$$F_t = w_F S_t / \bar{S}_t + (1-w_F) F_{t-L} \qquad (15\text{-}8)$$

Formula (15-8) has the same basic form as formula (15-1). If daily forecasts are made and if the cycle length is one week then L equals seven. The weight w_F lies between zero and unity, S_t is actual sales in the current period and $\bar{S}_t$ is the permanent component of sales in the current period.

Formula (15-8) is based on the permanent component of sales $\bar{S}_t$. Hence, F_t cannot be calculated until $\bar{S}_t$ has been determined. It is found by a modification of formula (15-1) shown below,

$$\bar{S}_t = w_e S_t / F_{t-L} + (1-w_e) \bar{S}_{t-1}. \qquad (15\text{-}9)$$

Using the permanent component $\bar{S}_t$ found by formula (15-9) and the seasonal permanent component F_{t-L+T} previously found by formula (15-8) the forecast for T periods into the future becomes,

$$S_{t,T} = \bar{S}_t F_{t-L+T}. \tag{15-10}$$

The forecast in formula (15-10) is of course based on sales information available up to period t and seasonality of sales information available up to period t-L+T which must be stored for each period during the length of the cycle. The forecast for sales in the next period, T=1, would thus be $S_{t,1}$ which equals $\bar{S}_t F_{t-L+1}$. If forecasts are to be made over a period exceeding the length of the cycle, i.e. if T>L then the forecast can still be made but the seasonal or cyclical weights, the F_t's must be used over again.

As can be observed from formula (15-10) the above seasonal or cyclical forecast model is a multiplicative model. Later in this chapter an additive seasonal or cyclical model will be discussed.

Associated with applying a seasonal or cyclical model is the problem of establishing a set of stable F_t's or also called seasonals or cyclicals. With a short cycle of a week or a month this is not too difficult. However, with an annual cycle each year only adds one observation to each seasonal and hence, several years of historical data must be available before the model can be expected to become a reasonable forecaster.

Trend Effects

Mechanical or objective forecasters such as discussed in this chapter should be able to respond to longer term upward or downward changes in sales. This can be accomplished by adding to the basic exponential model a permanent trend component, A_t, which is updated each period as follows,

$$A_t = w_A(\bar{S}_t - \bar{S}_{t-1}) + (1-w_A) A_{t-1} \qquad (15\text{-}11)$$

Formula (15-11) is of the same basic form as formula (15-1) and (15-8). The weight w_A lies between zero and unity.

The forecast of sales, based on consideration of a trend effect only, then becomes,

$$\bar{S}_t = w_e S_t + (1-w_e)(\bar{S}_{t-1} + A_{t-1}). \qquad (15\text{-}12)$$

Note that formula (15-12) adds the magnitude of the trend to the permanent component. It is therefore called an additive trend model. A multiplicative trend model will be discussed later in this chapter.

A forecast for T periods into the future is found by,

$$S_{t,T} = \bar{S}_t + TA_t. \qquad (15\text{-}13)$$

For T=1 the forecast $S_{t,1}$ would thus be simply the sum of $\bar{S}_t$ and of A_t.

The trend A_t is updated as frequently as the permanent component $\bar{S}_t$. Hence, only a limited amount of historical data is required to put the trend effect into operation.

EXPONENTIAL SMOOTHING EXTENSIONS[6]

The basic exponential approach to forecasting introduced above can be extended to several new types of stochastic processes. The ease of these extensions and their flexibility in handling different forecasting problems illustrates several attractive features of this forecasting approach.

A total of nine possible forecast models will be presented. Graphical illustrations implicitly suggest which model is the most appropriate to use on the basis of the historical sales generating process.

The Nine Exponential Models

The nine possible models mentioned above consist of three sets of three models. The first set, portrayed in Figure 15-1, consists of the basic exponential model,

---

[6] This section is based on: C.C. Pegels, "Exponential Forecasting-Some New Variations," Management Science, Vol. 15, No. 5 (January 1969), pp. 311-15.

Model 1-A, the basic model with a stable additive trend
effect, Model 1-B, and the basic model with a stable
multiplicative trend effect, Model 1-C. The second set
of three models, Models 2-A, 2-B, and 2-C, presented in
Figure 15-2, consists of the three models in Figure 15-1
with a stable additive seasonal effect superimposed on
them. The third set of three models, Models 3-A, 3-B, and
3-C, presented in Figure 15-3, again consists of the three
models in Figure 15-1, but now with a stable multiplicative
seasonal effect superimposed on them. Holt, et.al. only
discussed Models 1-A, 1-B, and 3-B, because they felt that
it is more generally useful to work with these models.
However, some of the nine models presented above, appear
as probable, and some even appear more probable in real-
life applications than the models with only additive trend
and multiplicative seasonals.

The nine models can be summarized by the formula,

$$\bar{S}_t = w_e d_{S,t} + (1-w_e)d_{T,t},$$

where $\bar{S}_t$ is the basic permanent forecast component in
period t, w_e is a weight, $0 \leq w_e \leq 1$, and $d_{S,t}$ and $d_{T,t}$
are given in Table 15-1. Each entry in the 3 x 3 matrix
of the table is directly related to the models shown in
Figures 15-1, 15-2, and 15-3. For instance, the model
with no trend effect but with an additive seasonal effect

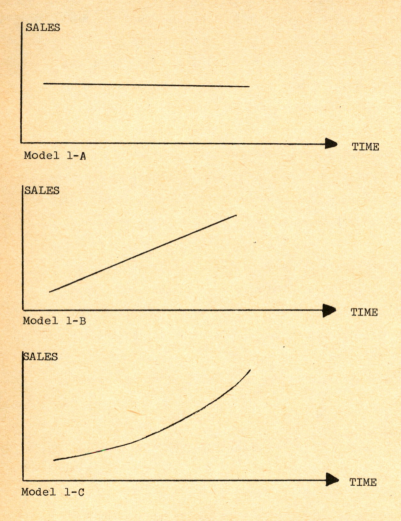

Figure 15-1

Forecasting Models Without Seasonal Effect

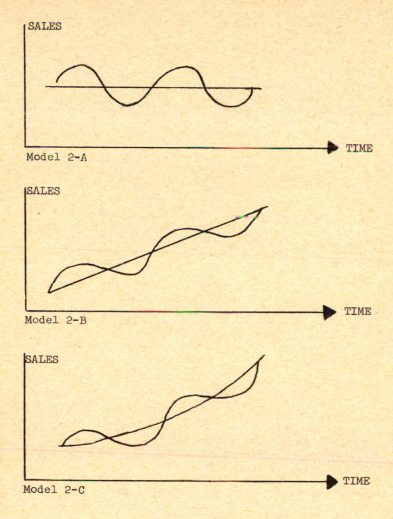

Figure 15-2

Forecasting Models With Additive Seasonal Effects

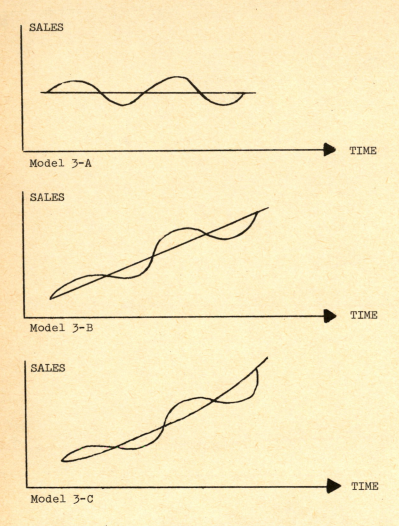

Figure 15-3

Forecasting Models With Multiplicative Seasonal Effect

is Model 2-A, obtained from row A and column 2.

The symbols denoted by $d_{S,t}$ and $d_{T,t}$ will now be discussed. Although some of the formulas were presented earlier in this chapter, they will be repeated here for continuity. The first one, S_t, is the actual sales in period t. The additive trend component, A_t, is derived from the formula:

$$A_t = w_A(\bar{S}_t - \bar{S}_{t-1}) + (1-w_A)A_{t-1},$$

and represents the incremental increase in the basic permanent forecast component for time period, t. w_A is a weight, $0 \leq w_A \leq 1$.

Table 15-1

Symbols Used in Summary Formula

		1 no seasonal effect	2 additive seasonal effect	3 multiplicative seasonal effect
A. no trend effect	$d_{S,t}$	S_t	$S_t + G_{t-L}$	S_t / F_{t-L}
	$d_{T,t}$	$\bar{S}_{t-1}$	$\bar{S}_{t-1}$	$\bar{S}_{t-1}$
B. additive trend effect	$d_{S,t}$	S_t	$S_t + G_{t-L}$	S_t / F_{t-L}
	$d_{T,t}$	$\bar{S}_{t-1} + A_{t-1}$	$\bar{S}_{t-1} + A_{t-1}$	$\bar{S}_{t-1} + A_{t-1}$
C. multiplicative trend effect	$d_{S,t}$	S_t	$S_t + G_{t-L}$	S_t / F_{t-L}
	$d_{T,t}$	$\bar{S}_{t-1} B_{t-1}$	$\bar{S}_{t-1} B_{t-1}$	$\bar{S}_{t-1} B_{t-1}$

The multiplicative trend component, B_t, is derived from the formula,

$$B_t = w_B \bar{S}_t / \bar{S}_{t-1} + (1-w_B)B_{t-1},$$

and represents the incremental increase in the basic sales forecast for time period, t. w_B is a weight, $0 \leq w_B \leq 1$.

The additive seasonal component, G_t, is derived from the formula,

$$G_t = w_G(\bar{S}_t - S_t) + (1-w_G)G_{t-L},$$

and represents the parameter which normalizes actual sales in the basic forecasting equation. w_G is a weight, $0 \leq w_G \leq 1$.

The multiplicative seasonal component, F_t, is derived from the formula,

$$F_t = w_F S_t / \bar{S}_t + (1-w_F)F_{t-L},$$

and represents the parameter which normalizes actual sales in the basic forecasting equation. w_F is a weight, $0 \leq w_F \leq 1$.

Actual sales forecasts for period t+T can now be obtained from the general formula,

$$S_{t,T} = g_t$$

where the g_t's for each model are the entries in Table 15-2.

Table 15-2

Sales Forecasts for the Nine Models

	1	2	3
	no seasonal effect	additive seasonal effect	multiplicative seasonal effect
A. no trend effect	$\bar{S}_t$	$\bar{S}_t + G_{t-L+T}$	$\bar{S}_t F_{t-L+T}$
B. additive effect	$\bar{S}_t + TA_t$	$\bar{S}_t + TA_t + G_{t-L+T}$	$(\bar{S}_t + TA_t) F_{t-L+T}$
C. multiplicative trend effect	$\bar{S}_t B_t^T$	$\bar{S}_t B_t^T + G_{t-L+T}$	$\bar{S}_t F_{t-L+T} B_t^T$

Comments on Models

In multi-item demand forecasting situations it may be necessary to prepare sales forecasts for tens of thousands of items every month. The models presented in this chapter, together with the appropriate computing machine, will perform this task accurately and efficiently. The graphical representatives of each proposed model even allow the selection of the appropriate model provided some information on historical sales or, in case of a new product, estimates of future sales behavior are known.

APPLICATION PROCEDURE

The exponential forecasting system discussed above
is rather easy to apply if the proper sequence of steps
are taken. A model with multiplicative seasonal and
additive trend will be used as the vehicle to illustrate
the procedure.

Determining the Starting Values

The first step is the determination of the length
of the cycle. Assume that the cycle is one week and forecasts are made for each day. Hence, a total of seven
periods constitute a cycle (L equals seven). Historical
sales over a period of five weeks have been collected and
are shown in Table 15-3. The historical sales figures are
shown in matrix format with the columns forming the five
weeks and the rows forming the seven days of the week.
Hence column averages provide average daily sales during
each of the five weeks and row averages provide average
daily sales for each specific day of the week. The day
or row averages are shown in the far right column and the
column averages appear in the bottom row.

Given the above information how can the exponential
model be applied? Assuming that the model will also be
tested out, although this is not a requirement, the
suggested beginning value for the permanent component

Table 15-3

Historical Sales Data by Day and by Week

| Day of Week | Week | | | | | Day Average in |
	1	2	3	4	5	Average Week
Sunday (1)	40	35	45	50	55	45
Monday (2)	320	325	330	340	360	335
Tuesday (3)	270	260	280	280	310	280
Wednesday (4)	410	420	415	410	445	420
Thursday (5)	350	340	370	360	380	360
Friday (6)	280	270	300	310	290	290
Saturday (7)	60	60	50	60	70	60
Week Average	247	244	256	259	273	256

$\bar{S}_t$ is the first week's average daily value of 247. The
beginning trend component, A_t, can be found by taking the
difference between the first week's daily average sales
and the last (fifth) week's daily average sales and dividing
the difference by the number of days in the four-week span.
This approach would result in a beginning value for A_t of
0.93 found by subtracting 273 from 247 and dividing the
difference by 28. To find the cyclical permanent component, F_t, it is suggested to take the ratio of each
specific day average (in the last column) to the average
for the average week. An alternate way is to take any one
of the previous weeks or the cumulative average of several

of the previous weeks. In Table 15-4 the F_t's based on Table 15-3's last column, the first week (first column), and the fifth week (second last column) are presented for comparison. It is interesting to note that they are quite similar. However, not all data will be as stable as that presented in this hypothetical example.

Table 15-4

Comparison of Several Seasonals

Day of Week	First Week Only	Fifth Week Only	Average of Five Weeks
Sunday	0.162	0.202	0.177
Monday	1.296	1.321	1.312
Tuesday	1.091	1.136	1.093
Wednesday	1.661	1.630	1.642
Thursday	1.419	1.393	1.409
Friday	1.132	1.062	1.132
Saturday	0.243	0.256	0.235
Sum	7.000	7.000	7.000

Applying the Method in Practice

When using a model with seasonal and upward or downward trend effects the problem arises in what sequence the three permanent components, $\bar{S}_t$, F_t or G_t and A_t or B_t should be selected. Since both the trend and seasonal permanent components are based on $\bar{S}_t$, it is of course required that

$\bar{S}_t$ is calculated first. However, to calculate $\bar{S}_t$ requires that the value of actual sales S_t is known. Following the $\bar{S}_t$ calculation either the trend or seasonal component can be determined using $\bar{S}_t$ and S_t as input. Finally the sales forecast or forecasts for T periods into the future, $S_{t,T}$ can be determined using $\bar{S}_t$, A_t and the appropriate F_t.

Selection of Weights

Determining the weights, w_e, w_A, w_B, w_F and w_G, can be done by the arbitrary method discussed before or if historical data are available by an error minimization procedure. However, the optimal weights determined by the error minimization procedure are based on historical data and therefore only optimal if the sales generation process remains unchanged.

The forecast error for period t+1 will be defined as the difference between actual sales in period t+1 and the sales forecast for period t+1, that is,

$$\epsilon_{t,1} = S_{t+1} - S_{t,1}$$

Note that $S_{t,1}$ is the forecast for period t+1 made in period t. In general the forecast for period t+T will generate the error,

$$\epsilon_{t,T} = S_{t+T} - S_{t,T}$$

As stated above the optimal weights will be defined as those that minimize the forecast error or rather the sum of squares of the forecast error. An adequate substitute for the sum of squares of the error is the variance of the error, which is estimated by,

$$\sigma_T^2 = \sum_{t=1}^{N} \epsilon_{t,T}^2 / (N-1)$$

for period t+T, where t = 1,2,..., n. Of course the variance can be estimated for each value of T. However, one would expect that the most critical forecast is the one for the next period, that is for period T=1. In that case the variance becomes,

$$\sigma_1^2 = \sum_{t=1}^{N} \epsilon_{t,1}^2 / (N-1)$$

To apply the method, of course, requires that each triple set $\{w_e, w_A, w_F\}$ is evaluated. Assuming that the weights are evaluated in increments of 0.1 from 0 to 1.0 results in an evaluation process of 11 x 11 x 11 = 1331 points. If this procedure is required for only a limited number of products then it is rather straightforward and manageable. For a large number of products it may become rather expensive and other methods should be investigated.

One of these is the method of steepest descent which Holt, et.al.[7] claim to be quite expensive. In general it will be found that the sensitivity of the selected values for the weights is not very significant in the range that is normally found optimal. This range usually covers the values from 0.05 to 0.25.

CONTROL CHARTING SALES FORECASTS

An effective tool for keeping track of forecast errors is the statistical control chart which is extensively used in quality control. Controlling forecast errors is, however, considerably more difficult than controlling quality. In quality control applications a sample is drawn and the sample mean of the variable under control is plotted on the control chart. Since sample means are approximately normally distributed, the control limits can be accurately set for specified confidence intervals.

In the control of forecast errors it is not so easy to use and plot sample means, and therefore the difficult problem of setting meaningful control limits must be faced. It is possible, of course, to assume that forecast errors are normally distributed, but there usually is no assurance that the assumption is in agreement with reality.

---

[7] Holt, et.al., op. cit., p.264.

Therefore a compromise will be proposed. In the following example it will not be assumed that individual observations of forecast errors are normally distributed, but the sample means of four consecutive weekly forecast errors will be assumed to be normally distributed. On the basis of this assumption control limits will be set for a given confidence interval and a total of thirteen observation points during a one-year interval will thus be obtained.

Suppose the following weekly forecast error data have been collected over a period of two years. The first year will be used to determine the control limits based on a specified confidence interval, and the second year sample means will be plotted on the control chart. The forecast error data for the two-year period are shown in Table 15-5, and the control chart is shown in Figure 15-4.

Based on the first year data a mean of 4.6 is obtained which indicates that during the first year the average forecast process had been somewhat optimistic. The sample standard deviation was obtained by the formula,

$$S_{\bar{x}} = \left[\sum_{i=1}^{n} (\bar{x}_i - \bar{\bar{x}})^2 / (n-1) \right]^{\frac{1}{2}}$$

where $\bar{x}_i$ is the sample mean for samples of size four, n is the number of samples (n=13) used to derive the sample standard deviation, ($s_{\bar{x}} = 9.4$) and $\bar{\bar{x}}$ is the mean for the one-year period.

Table 15-5

Forecast Error Data

First Year				Second Year			
Week	Error	Week	Error	Week	Error	Week	Error
1	+49	27	-7	1	+5	27	+5
2	+1	28	-1	2	+32	28	-3
3	+1	29	+3	3	+33	29	-5
4	+7	30	+3	4	-21	30	+5
5	0	31	-45	5	+38	31	-1
6	-8	32	+2	6	+32	32	+9
7	-10	33	+3	7	-8	33	-3
8	-42	34	0	8	+5	34	+28
9	-4	35	+12	9	+19	35	-7
10	-1	36	-7	10	-9	36	-1
11	+8	37	+48	11	+8	37	+5
12	-9	38	0	12	+5	38	+41
13	+13	39	+7	13	-10	39	0
14	+2	40	+1	14	+2	40	-28
15	+37	41	+4	15	-4	41	+1
16	-6	42	-4	16	+8	42	+44
17	+22	43	+2	17	-7	43	-9
18	+1	44	-35	18	-4	44	+7
19	-1	45	-14	19	-1	45	+1
20	+19	46	-5	20	-28	46	+47
21	+6	47	-9	21	0	47	0
22	-1	48	+2	22	+22	48	+5
23	+6	49	+7	23	-14	49	+7
24	0	50	+1	24	+2	50	+4
25	+43	51	+5	25	+9	51	-48
26	-8	52	+8	26	0	52	+11

384

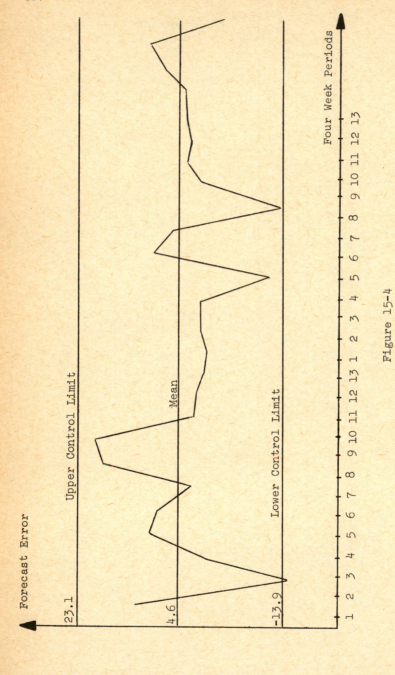

Figure 15-4

Control Chart for Forecast Errors

Based on a 95% confidence interval the control limits were set at 23.1 for the upper control limit and -13.9 for the lower control limit.

The reader may wonder why the mean of the forecast error is not closer to zero. This questioning is of course appropriate but if a specific forecast model is used which generates the error as indicated and if the order generating process has not changed then the control chart shown is the appropriate one. Another issue, unrelated to the control chart, is the question if the forecast model is sufficiently accurate. If it is decided it is not then of course it should be modified or improved and a new control chart should then be developed for the new forecasting model.

EXERCISES

1. Using the data in Table 15-1, find exponential smoothing parameters such as a starting trend component A_t, a set of starting seasonal components F_t's, and a starting permanent component $\bar{S}_t$.

2. Apply exponential smoothing model with components developed in problem 1 to the set of two-year data in Table 15-1. Use a value of 0.10 for the weights w_e, w_A and w_F. Compare forecasts $S_{t,1}$ with actual sales.

3. Using the data in Table 15-3, find the exponential smoothing parameters identified in the text as B_t and G_t.

4. Apply exponential smoothing model with a selected set of trend and seasonal effects (i.e. use A_t or B_t and F_t or G_t) and make forecasts $S_{t,1}$ for the five weeks indicated in Table 15-3. Use values of 0.10 for weights.

5. Using the error minimization model find optimum set of weights for the model you selected in problem 4.

6. Prove that exponential smoothing results in an un-biased average--that is--the sum of the coefficients equals one.

7. A product is in its period of rapid sales expansion.

While forecasting the short run demand, the use of a
weighted moving average would be most meaningful.
Discuss.

8. Redding distributors was a leading national distributor
of plumbing and hardware supplies. The company's
product line of over 3,000 items was sold through
95 branches in 40 states served by 4 regional warehouses.
Company management had reached the conclusion that in
recent years Redding's inventory management practices
and policies had become inadequate. As a step toward
changing this situation, a new position for inventory
management planning was created in the home office.
The position was filled by Mr. Gene Dougall, a recent
graduate of a well-known business school.
Redding for many years stressed service in its distribution
policies. Each regional warehouse carried the complete
line of products. The branches carried limited quantities
of specific items in inventory depending on their location, size and the extent of their market area.
Orders from branches to regional warehouses were shipped
within 24 hours of receipt. Orders sent from the warehouses to manufacturing sources of the company's
product line often were not received for 1 to 3 months.
Each warehouse operated nearly independently of the
others. Occasional inter-warehouse shipments took

place but these were avoided since shortages had resulted in the past at the warehouse from which such
shipments were made.

Lower inventory costs could be achieved by reducing
inventory. However, the plumbing and hardware supply
field was extremely competitive and it was Redding's
service policy that the regional warehouses have sufficient
stock on hand to fill branch orders on a unit basis
within 24 hours at least 95% of the time. For instance,
if 4 orders for 1,000 units each were received, a 95%
service level on a unit basis would be provided by
shipment of 950 units for each order.

Despite the company's service objectives, Mr. Frank
Summers, Vice-President of Warehouse Operations to whom
Mr. Dougall was assigned to report, stated that inventory
costs -- including: warehouse space, obsolescence,
working capital invested in iventors, taxes, insurance,
warehouse stock handling and spoilage -- was an area
of company operations where costs could and should be
reduced.

After familiarizing himself with company operations, Mr.
Dougall decided to attempt to analyze the behavior of
a number of warehouse items. He segregated items by
several measures, including: unit cost, volume, space
requirements, speed of obsolescence, and perishability.
Two items selected for analysis were high-volume, low-cost

389

products. Actual monthly demand for these items for
the past 36 months, as reported by one of the warehouses
was as follows:

	Part Number				Part Number	
Month	29	34		Month	29	34
1	4900	6400		19	3900	5700
2	4600	7100		20	6800	5300
3	3900	6900		21	4900	5100
4	4100	6100		22	4500	6300
5	6200	6200		23	4100	6800
6	4200	6300		24	5100	6500
7	4300	5900		25	6200	6300
8	3600	6600		26	4100	6900
9	3800	7100		27	3900	7100
10	4500	6900		28	3900	6400
11	3900	6100		29	4100	6100
12	4500	5800		30	4700	5900
13	4600	5500		31	3700	6300
14	3100	6300		32	4500	6100
15	6100	6500		33	4900	5800
16	4800	4900		34	6000	6900
17	4600	6900		35	4500	6400
18	4100	7200		36	4400	6300

The home office sales department reported that future
demand for these two items was expected to follow the

historical pattern. Mr. Dougall decided to use the historical data to evaluate what inventory management practice within the bounds of company service policy would have achieved the lowest inventory costs.

9. Sales of new automobiles by the Smith-Hudson Agency have been recorded for the past six years by quarterly periods as shown below.

Year	Quarter	Sales of Automobiles
1965	1	164
	2	289
	3	215
	4	186
1966	1	215
	2	373
	3	265
	4	192
1967	1	191
	2	345
	3	253
	4	200
1968	1	217
	2	411
	3	289
	4	253
1969	1	230
	2	368
	3	264
	4	226
1970	1	246
	2	458
	3	367
	4	360

Using the moving average method develop a forecast model considering trend and seasonal effects.

10. Using the data in exercise 9 find exponential smoothing parameters such as starting trend component A_t, a set of starting seasonal components F_t's, and a starting permanent component $\overline{S}_t$.

11. The three weights used in this chapter are w_e for the standard component, w_A or w_B for the trend component and w_F or w_G for the seasonal or cyclical component. Would you expect the values of these weights to be equal if all were used in the same model? Assuming that they are different how would you expect the three weights to differ in value? For instance $w_e \leq w_A \leq w_F$, or $w_A \leq w_e \leq w_F$, etc.

12. How is n in the moving average forecasting model related to w_e in the exponential smoothing model? Does knowledge of the one aid in determining the other?

13. What is the major disadvantage of moving average or exponential smoothing forecasting methods in comparison with the statistical forecasting method discussed in the previous chapter?

14. Discount Department Stores actual sales over the past two years compare with forecasted sales as shown in the table below. Prepare a sales control chart on the basis of 1969 sales and plot the 1970 sales on this chart. Would you want to revise your chart at the end of 1970 or not?

| | Sales in Millions of Dollars | | | |
Month	Forecasted (69)	Actual (69)	Forecasted (70)	Actual (70)
January	79	81	86	87
February	78	80	85	86
March	90	94	97	102
April	101	110	108	113
May	100	108	109	114
June	101	103	108	116
July	89	94	97	101
August	92	102	106	114
September	107	115	115	122
October	112	111	122	117
November	128	139	139	143
December	194	209	205	231

15. The following table shows historical demand data for demand of widgets.

Month	1968	1969	1970
January	111	134	87
February	90	152	102
March	132	59	216
April	89	228	139
May	113	78	109
June	92	120	85
July	61	175	119
August	154	52	196
September	107	149	120
October	138	104	238
November	114	32	147
December	166	159	176

Using the data for 1968 and 1969 develop values for the initial conditions required by an exponential smoothing model with an additive trend component (no seasonal). Let $w_e = w_A = 0.1$. Make a forecast for 1970 and compute the forecast error for each month. Then calculate the standard deviation of forecast errors.

16. Using the data in exercise 15 forecast 1970 demand using a moving average forecast method based on 1968 and 1969 historical demand data. Calculate the standard deviation of forecast errors and compare the standard deviation with exercise 15. Which method of forecasting would you select?

17. Production requirements for Meridian Modems over the past 36 months consisted of the following amounts.

	Production Requirements		
Month	1968	1969	1970
January	10,000	9,000	10,000
February	8,000	9,000	10,000
March	4,000	5,000	5,500
April	3,000	4,000	4,500
May	3,000	3,000	2,500
June	2,500	3,000	2,000
July	4,000	5,000	3,000
August	6,000	6,000	5,000
September	8,000	9,000	10,000
October	12,000	11,000	12,000
November	15,000	14,000	16,000
December	12,000	13,000	13,000

Forecast the 1970 requirements using 1968 and 1969 to develop an exponential smoothing model with additive trend and multiplicative seasonal effects. Use weights of 0.1 for the three weights.

18. Calculate the error between the 1970 forecasts and the 1970 actual requirements for exercise 17. Also find the standard deviation of the errors.

CHAPTER 16.--DETERMINATION OF DECISION RULES FOR PRODUCTION, INVENTORIES AND WORK FORCE LEVELS

How should production and employment levels be adjusted to fluctuations in sales? This question was usually resolved by adjusting the work force in the period prior to the advent of unemployment insurance, severance pay, and guaranteed annual wage. For most firms today, changing the size of the work force has become a considerable cost item and alternative means of responding to changes in sales levels must be found.

INTRODUCTION

In this chapter the problem of setting the production rate and size of the work force will be explored. Linear decision rules will be derived by classical calculus techniques from cost functions that are or can be approximated by quadratic cost functions.

Two models will be discussed and one case will be evaluated to determine the effect that the use of linear decision rules can have on a firm's production and work force planning decisions. The first model applies to a general firm that maintains some inventory and has a varying and seasonal demand; the other model applies to a job shop operation that does not and essentially cannot maintain inventory and has a highly fluctuating seasonal

demand. The Holt et.at.[1] approach represents the first
model and a difference equation approach developed by the
author[2] will be used on the job shop operation.

LINEAR DECISION RULES

The linear decision rules will determine the production rate and the size of the work force for future
periods. These rules are obtained from a model consisting
of a number of quadratic cost functions for overtime cost,
inventory cost, shortage cost and lay-off and hiring cost.
The forecast for future sales, current inventory level and
current work force are the inputs to the model, and hence
accurate sales forecasts will improve the quality of the
decision rules.

The decision rules are optimal in the sense that they
minimize all costs associated with a response to sales
fluctuations. It may thus not necessarily be the rule management may always want to follow. However, management
can obtain a quantitative estimate how much a policy change
from the optimal method of operation costs.

---

[1]C.C. Holt, Franco Modigliani, J.F. Muth and H.A. Simon.
Planning Production, Inventories and Work Force, Englewood
Cliffs, N.J.: Prentice-Hall, Inc., 1960, pp. 47-121.

[2]C.C. Pegels, "Work Force Planning for the Job Shop,"
Logistics Review, Vol. 5, No. 21, pp. 41-48.

Payroll Cost

Before discussing the two models in more detail a
short review of the various cost functions that are relevant to the two models will be made. The basic one,
regular payroll cost, excluding overtime cost is presented
in Figure 16-1. As the work force increases, the payroll
cost increases. Payroll cost can be expressed algebraically
as,

$$C_1 W_t$$

where W_t is size of work force in period t and C_1 is a constant equal to the monthly salary if the period equals a
month and if W_t is measured in workers.

Hiring and Layoff Costs

The next cost function that will be considered is
hiring and lay-off cost. Costs in this case are not
associated with the size of the work force but rather with
changes in its size. Figure 16-2 gives a representation of
this cost function. The cost of hiring and laying off
workers rises with the number hired and the number laid off.
Whether this rise is linear or quadratic may be open to
debate, but it is reasonable to approximate a linear cost
function with a quadratic cost function for a certain region of the function. Hiring and lay-off costs can be

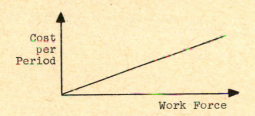

Figure 16-1

Payroll Cost Function Excluding Overtime

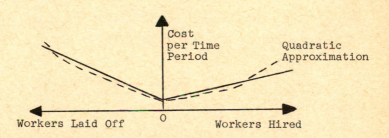

Figure 16-2

Hiring and Layoff Cost Functions

expressed algebraically as,

$$C_2(W_t-W_{t-1}-C_{11})^2,$$

where C_2 and C_{11} are empirically-based constants.

Overtime and Idle Time Costs

The overtime cost function is more difficult to determine. It is not based on the size of the work force or on the production rate but rather on the production rate given the work force. Overtime cost may also vary for amount of overtime worked. The overtime rate may be 50% of regular pay for weekdays and 100% of regular pay for weekends. Hence a quadratic approximation may in most cases be more accurate than a linear estimate. For a given work force the relation is shown in Figure 16-3. Any idle time incurred is included in the payroll cost function discussed before.

The algebraic expression for the quadratic approximation of overtime is,

$$C_3(P_t-C_4W_t)^2 + C_{12}P_tW_t. \qquad (16-1)$$

As production, P_t exceeds C_4W_t, a level set by the size of the work force overtime costs increase. The coefficients C_3, C_4 and C_{12} are empirically-based. The second term is added to improve the approximation but may be deleted if

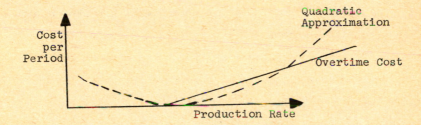

Figure 16-3

Overtime Cost Function

not found necessary. Since the overtime cost curve is
based on a given work force W_t, there is a whole family
of cost curves for each conceivable value of W_t. They
are, however, all expressed by (16-1).

Inventory and Shortage Costs

Inventory and shortage costs depend on the inventory
on hand and on the orders that could not be filled. As
the inventory varies from some optimal (minimum cost)
level, the inventory holding and shortage cost also varies.
The cost relation is shown in Figure 16-4. A quadratic
approximation is shown in dotted lines. The minimum cost
in the example does not occur at zero inventory but at some
positive value. This is not necessarily so in every case.
It is also conceivable that the cost curve attains a minimum
at some negative net inventory value.

The algebraic expression for the quadratic approximation of inventory holding and shortage cost is,

$$c_7(I_t - c_8 - c_9 S_t)^2$$

The term $c_9 S_t$ moves the cost curve along the horizontal
axis, its location dependent on the level of sales in
period t. The net inventory, I_t, in period t is the
amount of inventory or shortages at the end of the period.
Empirically-based coefficients are c_7, c_8 and c_9.

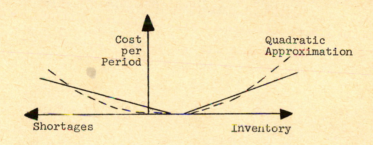

Figure 16-4

Inventory and Shortage Cost Functions

GENERAL MODEL

The four cost functions discussed above can now be summed to obtain the total cost function for period t. Total cost is,

$$C_t = C_1 W_t \quad \text{(Regular Payroll Costs)}$$
$$+ C_2(W_t - W_{t-1} - C_{11})^2 \quad \text{(Hiring and Layoff Costs)}$$
$$+ C_3(P_t - C_4 W_t)^2 + C_{12}P_t W_t \quad \text{(Overtime Costs)}$$
$$+ C_7(I_t - C_8 - C_9 S_t)^2 \quad \text{(Inventory Connected Costs)}$$

The decision problem can now be stated formally. Find the production level and work force level decisions that minimize $E(C_T)$, where

$$C_T = \sum_{t=1}^{T} C_t \tag{16-2}$$

subjects to the constraints,

$$I_{t-1} = I_t + S_t - P_t \quad t = 1,2,\ldots,T \tag{16-3}$$

Derivation of Decision Rules

The optimal production and employment decisions are those that minimize the expected value of total cost, C_T. This cost is the sum of the costs attributable to T periods or months as shown by (16-2) The relation between inventory at the beginning of each month, production during the month, sales during the month, and inventory at the end of the month is shown by (16-3).

The decision rules may now be obtained by differentiating with respect to each decision variable. A set of linear equations is then obtained and further mathematical manipulation provides the decision rules.

Before analyzing the decision rules, the rules will be presented and briefly discussed. One decision rule sets the production rate, and another determines the work force. The production rule and work force rule both incorporate a weighted average of the forecasts of future orders; since the forecasts of future orders are averaged, production and work force are smoothed, so that there is an optimal response to the fluctuations of forecasted orders. The weight given to future orders declines rapidly as the forecast extends farther into the future. This makes sense, intuitively, because it is usually not efficient to produce currently for shipment in the distant future.

Decision Rules for Production and Work Force

The production decision rule for period t is,

$$P_t = \sum_{r=0}^{R} v_r S_{t+r} + k_1 W_{t-1} + k_2 - k_3 I_{t-1}$$

and the work force decision rule for period t is,

$$W_t = \sum_{r=0}^{R} w_r S_{t+r} + k_4 W_{t-1} + k_5 - k_6 I_{t-1}$$

where v_r and w_r are the weights for each future period

k_i, i=1,..., 6 are constants determined by the results of the derivation.

R is the number of future periods of sales forecasts to be used for the decision rule.

P_t is the number of units of product that should be produced during the next period, t.

W_t is the size of the work force (number of employees) that will be required for the next period, t.

W_{t-1} is the size of the work force at the end of the present period, t-1.

W_t-W_{t-1} is the number of employees that should be hired, or if negative the number of employees that should be laid off.

S_t, S_{t+r} are the forecasts of number of units of product that will be sold during period, t, t+r.

I_{t-1} is the number of units of inventory minus the number of units on back order at the beginning of period t or at the end of period t-1.

Analysis of Decision Rules

Analysis of the two decision rules reveals some interesting features. Both rules depend on exactly the same variables, the only difference being the constants and the weights for future orders. Both rules are a function of work force in the previous period, W_{t-1}, inventory in the previous period, I_{t-1}, and unit sales in future periods plus a constant.

There is a fairly complex interaction between the two decision rules. The production of one period affects the net inventory position at the end of the period. This in turn influences the employment decision in the second period which then influences the production decision in the third period. Hence there is a continual dynamic interaction between the two decisions.

The weights that are applied to the sales forecasts and the feedback factors in the two decision rules determine the production and employment responses to fluctuations of orders and thereby indicate how much of these fluctuations should be absorbed by work force fluctuations, overtime fluctuations, and inventory and back order fluctuations in order to minimize costs.

THE JOB SHOP OPERATION

In the case of the typical job shop operation, production for inventory is not feasible and hence inventory cannot be used to smooth the often drastic work force level changes. The only way to smooth work force levels is by the judicious use of overtime or idle time. The proposed model to be discussed next will provide the means to minimize total labor-related costs for the job shop operation.

The proposed model is a modified Holt, et.al. model, which was developed after it was determined that the

Holt, et.al. model could not be applied to a job shop operation. The proposed model evolves into a set of difference
equations which, when solved simultaneously, provide the
linear decision rules for the minimum cost work force level
in the next production period.

The Cost Model for the Job Shop Operation

The cost function to be minimized consists of five
cost components, payroll cost excluding overtime premium
and idle time cost, hiring cost, layoff cost, overtime
premium cost, and idle time cost. The five cost items are
summed over the planning horizon which typically consists
of twelve months, where each month is a planning period.

The cost function components in this model are thus
somewhat different from the previous cost model. The first
term, payroll cost, excludes idle time which was included
in the previous model. The other difference in the payroll
cost term is overtime premium cost which is excluded in
the job shop model whereas all overtime cost was excluded
in the previous model. There is no change in hiring and
layoff costs. However, the last cost term in the job shop
cost model only includes overtime premium, instead of all
overtime cost, and also includes idle time cost which was
excluded in the previous model.

In mathematical terms the problem can be stated as:

$$\text{Minimize} \quad E(\cdot) = \sum_{t=1}^{T} [C_1 P_t / C_4 \quad \text{(payroll cost excluding idle time cost and overtime premium)}$$

$$+ C_2 (W_t - W_{t-1})^2 \text{ (hiring and layoff costs) (16-4)}$$

$$+ C_3 (P_t - C_4 W_t)^2] \text{ (overtime premium and idle time cost)} \quad \text{(16-5)}$$

$$\text{subject to} \quad S_t = P_t \quad \text{for } t = 1, \ldots, T, \quad \text{(16-6)}$$

where the C_j, $j = 1, \ldots, 4$ are empirically determined cost coefficients. S_t is the number of units[3] sold during period $t = 1$, and sales forecasts for periods $t = 2, \ldots, T$; similarly, P_t is the number of units produced during period $t = 1$, and number of units to be produced subject to (16-6) for periods $t = 2, \ldots, T$. W_t is the work force level for period $t = 1, \ldots, T$, and $W_t - W_{t-1}$ is the change in work force level from period $t-1$ to period t.

Note that (16-4) and (16-5) for each period are quadratic convex cost functions, each of which is an approximation to two linear or nonlinear cost functions.

---

[3]The term "unit" may be interpreted literally if it applies, for instance, number of barrels, refrigerators, bicycles, and so forth. If the production output is not in literal units, then the term "unit" may be interpreted as dollars of output or thousands of dollars of output, or also gallons of paint, and so forth.

An estimate of the coefficients for the two quadratic cost functions would be obtained by estimating the individual cost functions first, and then approximating the two individual cost functions with one quadratic convex cost function.

To obtain the work force decision rule differentiate E with respect to W_r, $r = 1,\ldots,T-1$ and obtain,

$$\frac{\partial E}{\partial W_r} = 2C_2(W_r - W_{r-1}) - 2C_2(W_{r+1} - W_r) - 2C_3C_4(P_r - C_4W_r)$$

$$r = 1,2,\ldots,T-1$$

and

$$\frac{\partial W_{t-1}}{\partial W_r} = \begin{cases} 1 & \text{if } t=r+1 \\ 0 & \text{otherwise} \end{cases}$$

and

$$\frac{\partial W_t}{\partial W_r} = \begin{cases} 1 & \text{if } t=r \\ 0 & \text{otherwise} \end{cases}$$

Since the total cost function is convex the minimum cost work force level can be found by letting

$$\frac{\partial E}{\partial W_r} = 0 \qquad r = t,\ t+1,\ t+2,\ldots,$$

which results in,

$$-2C_2W_{t+1} + 4C_2W_t - 2C_2W_{t-1} + 2C_3C_4^2W_t - 2C_3C_4P_t = 0$$

$$-2C_2W_{t+2} + 4C_2W_{t+1} - 2C_2W_t + 2C_3C_4^2W_{t+1} - 2C_3C_4P_{t+1} = 0$$

$$-2C_2W_{t+3} + 4C_2W_{t+2} - 2C_2W_{t+1} + 2C_3C_4^2W_{t+2} - 2C_3C_4P_{t+2} = 0$$

The above set of equations can be extended indefinitely. However, it will be shown that the three equations shown above are sufficient to derive the linear work force decision rules.

The sales or production forecasts P_t, P_{t+1}, P_{t+2}, and W_{t-1} are known[4] and an estimate of W_{t+3} may be used, because the coefficient derived for W_{t+3} is very small. Hence, the result is a set of three equations in three unknowns. These equations can be solved simultaneously for W_t, W_{t+1} and W_{t+2}. However, normally only the work force decision for the next period, that is period t, is required. The work force decision rule for period t then becomes,

$$W_t = 2C_2 C_{43} W_{t-1} + 2C_3 C_4 C_{43} P_t + \frac{4C_2 C_3 C_4}{C_{42}} P_{t+1} +$$

$$+ \frac{8C_2^2 C_3 C_4}{C_{41} C_{42}} P_{t+2} + \frac{8C_2^3}{C_{41} C_{42}} W_{t+3},$$

where $C_{41} = 4C_2 + 2C_3 C_4^2$,

$C_{42} = C_{41}^2 - 8C_2^2$, and

$$C_{43} = \frac{C_{41}^2 - 4C_2^2}{C_{41} C_{42}}$$

---

[4] P_t is actual sales for period t, and P_{t+1} and P_{t+2} are the sales forecasts for periods t+1 and t+2 respectively.

APPLICATION OF MODEL

The model was applied to a real job shop situation. The cost parameter estimates were, $C_1 = 500$, $C_2 = 5$, $C_3 = 50$, and $C_4 = 1.428$. Sales forecasts were available for the next three periods as shown in Table 16-1. The figures in the first column are actual sales and planned production levels.

Using the derived work force decision rule expected minimum cost work force levels were calculated as shown in Table 16-2. The decision rule with the derived coefficients is shown below,

$$W_t = .044752W_{t-1} + .639061P_t + .028599P_{t+1} +$$

$$+ .001277P_{t+2} + .000042W_{t+3}$$

Note that the coefficient for W_{t+3} is very small relative to the other coefficients. Therefore, the last component will have very little effect on the work force level. As a matter of fact, in the above example the last term could be dropped without affecting the work force level decisions.

In Table 16-2, a naive rule work force level is also shown for comparison with the decision rule work force level. The naive rule specifies the work force level required to produce in response to the actual sales without any overtime or idle time.

Table 16-1

Actual Sales and Sales Forecasts for Future Periods

Period	Actual Sales	Forecasted Sales		
t	S_t	S_{t+1}	S_{t+2}	S_{t+3}
1	236	200	367	654
2	160	295	589	791
3	224	669	844	747
4	685	771	755	896
5	782	718	870	838
6	668	797	758	640
7	788	751	642	694
8	709	570	618	489
9	553	635	488	378
10	685	540	418	258
11	538	407	250	197
12	391	246	195	292
13	243	244	314	544
14	273	366	567	661
15	400	623	722	756
16	539	655	692	787
17	662	640	726	654
18	671	725	655	598
19	720	729	665	673

Note: $S_t = P_t$

Table 16-2

Comparison of Optimum and Non-Optimum Work Force Level

Period t	Naive Rule Work Force Level	Decision Rule Work Force Level
0	270	-
1	165	169
2	112	119
3	157	169
4	479	468
5	547	542
6	468	475
7	551	547
8	496	494
9	387	394
10	479	471
11	377	377
12	274	274
13	170	175
14	191	193
15	280	283
16	377	377
17	463	459
18	470	470
19	504	503

Effectiveness of Model Application

In order to obtain a comparison between a minimum cost policy and a naive policy, a cost comparison is presented in Table 16-3. With the naive policy the number of people on the payroll are determined by the number of units to be produced. No overtime or idle time is planned at all. The cost saving by using the linear work force decision rule amounts to $56,385. This is only slightly larger than a 1% labor saving. However, in terms of profits or contribution to fixed cost, the cost saving would normally be substantial. For instance, if profit or contribution to fixed cost is 5 percent of labor cost the savings obtained through application of the work force decision rule would increase profit or contribution by approximately 25 percent.

Table 16-3

Cost Difference Between Optimal and Naive Rule

Period t	Naive Policy Cost	Optimal Policy Cost
1	$137,642	$135,322
2	69,194	73,444
3	88,446	107,021
4	751,510	699,105
5	297,231	303,256
6	264,771	262,061
7	309,969	303,244
8	263,027	262,397
9	251,676	248,356
10	280,915	275,205
11	240,131	232,341
12	189,758	189,808
13	138,010	136,420
14	197,454	197,524
15	179,465	181,610
16	235,506	232,641
17	268,448	266,888
18	234,860	235,270
19	257,528	257,243
Total	$4,655,541	$4,599,156

Cost saving, optimal policy vs. naive policy: $56,385.00

EXERCISES

1. How could payroll cost be expressed in an alternate algebraic form from the one given in the text if a fixed number of working foremen were employed regardless of the level of production? At lower levels of production these foremen would do a considerable amount of actual productive work. At higher levels of production, the foremen would be fully occupied with supervisory activities.

2. The production and work force planning model assumes that production can be measured in some aggregate form such as gallons of paints, barrels of oil, etc. How could the model be applied to a plant producing a product which is difficult to aggregate such as, (1) a paperbox plant producing thousands of different product lines; (2) a plant producing both cars and trucks; (3) a plant producing a wide variety of electronic components.

3. Calculate an efficient production plan for one year given the requirements shown below.

Initial Inventory	3 units
Production rate	1 unit/person/month
Regular time labor cost	$400/person/month
Overtime labor cost	$600/per unit (maximum of 2 per month)
Hiring cost	$200/person

Layoff cost	$500/person	
Carrying cost	$100/ unit/ month	
Desired closing Inventory	4 at end of December	
Labor force needed	1 unit/person/month	
Marginal revenue	$100/unit	

Month	Beginning Inventory	Sales Forecast	Work Force Level
January	3	10	7
February		6	
March		6	
April		7	
May		7	
June		9	
July		9	
August		8	
September		8	
October		6	
November		4	
December		6	

4. Typical Toy Manufacturing has a highly seasonal demand for its products. Demand usually peaks during November for December store sales and dips to its lowest level in May. Labor content of Typical's products is quite linearly related to the total product cost and as a result total sales volume in dollars is directly related to standard labor hours in a ratio of eight to one.

However, during peak demand periods, even with an expanded work force, large amounts of overtime still
must be worked at additional costs of fifty to one
hundred percent of regular labor cost. Because of the
nature of Typical's products production only takes
place in reponse to orders. Toys which are in favor
one month may be completely out of favor the next month
and therefore it is Typical's policy not to produce
anything for inventory. However, a minimum of one
to two months of lead time is common in the industry
and at the beginning of each month the sales level for
the current month is known. The hourly rate of production
workers is $3.00 per hour and does not vary much as
most production activities are of an unskilled to
semi-skilled nature. The cost of hiring and training
a worker costs approximately $300 for each worker hired.
Layoff costs, consisting of severance payments, vary
with the length of service of the respective worker
but have averaged $200 per worker over the past three
years. Average hourly overtime costs amounts to sixty
percent of regular time costs. Develop the optimum work
force and overtime planning model for Typical Toy
Manufacturing. Based on the following actual sales and
forecasted sales in thousands of dollars determine
optimum work force levels for the period from June 1970
to May 1971. Assume that inventory on hand at end of
May 1970 is negligible.

Month	Forecast Sales	Actual Sales	Actual work force level
January '70	80	76	150
February	85	79	150
March	60	57	130
April	40	39	120
May	25	26	100
June	35		
July	50		
August	75		
September	100		
October	125		
November	175		
December	75		
January '71	75		
February	80		
March	60		
April	40		
May	25		

5. What assumptions must be made regarding the make-up of the work force and production mixes to apply the models discussed in this chapter? Indicate in which cases simplifying assumptions are acceptable and in which cases simplifying assumptions can not be made.

6. Formulate the total cost function for work force, inventory and overtime costs if the payroll cost function is,

   $$C_1 W_t - C_{17} W_t^2$$

   Will more or less overtime be worked with this cost function? Will
   more or fewer people be laid off during a temporary dip in demand?

7. Develop the total cost function for the non-inventory work force
   planning model if payroll cost excluding idle time and overtime
   premium amount to

   $$C_1 P_t / C_4 - C_5 P_t^2$$

   Will there be any change in the amount of overtime worked? Will more
   or fewer people be laid off during a temporary dip in demand?

8. Suppose that the overtime premium and idle time cost function (16-5)
   in the non-inventory work force planning model is changed to

   $$C_3 (P_t - C_4 W_t + C_6)^2$$

   Will overtime worked increase or decrease?

   Will more or fewer people be laid off during a temporary dip in demand?

9. Find the minimum cost work force levels for the non-inventory work
   force planning model after the overtime and idle time cost function is
   changed to the one listed in exercise 8.

10. Apply the revised non-inventory work force planning model derived in
    exercise 9 to the problem listed in Table 16-1. Parameter C_6 equals 2.5.

CHAPTER 17.--PRODUCTION, INVENTORY AND WORK FORCE PLANNING
WITH LINEAR PROGRAMMING MODEL

In the previous chapter a production and employment
planning model was discussed which produced production,
employment and inventory decisions in the form of linear
decision rules. In this chapter similar cost functions
will be used as in the previous chapter, but instead of
using quadratic programming from which the linear decision
rules were derived the approach used in this chapter uses
linear programming as proposed by Hanssmann and Hess[1].

MODEL FORMULATION

The problem is stated as follows. Given the monthly
demands for the product produced by a manufacturing plant,
what should be the monthly production rates and work force
levels in order to minimize the total cost of regular payroll
and overtime, hiring and layoffs, inventory and shortages
incurred during a specified planning interval of several
months?

In the previous chapter the above problem was solved
by assuming that linear cost functions could be approximated

---

[1]Fred Hanssmann and S.W. Hess, "A Linear Programming
Approach to Production and Employment Scheduling," Management
Technology, January 1960, pp. 46-51.

by quadratic cost functions. In this chapter the cost functions will remain linear and if the linear assumption is a good representation of reality then solution methods using linear functions seem desirable.

The Cost Functions

It is assumed that the number of man-hours required to turn out a production quantity P_i

$$M_i = kP_i \qquad (17-1)$$

where k is a constant of proportionality and M_i is man-hours. If W_i is workforce level in man-hours, then in the case of $M_i > W_i$ an amount $M_i - W_i$ of overtime in overtime man-hours is required.

For any real number a the following will be defined:

$$a^+ = \begin{cases} |a| & \text{for } a \geq 0 \\ 0 & \text{otherwise} \end{cases} \qquad (17-2)$$

and

$$a^- = \begin{cases} 0 & \text{for } a \geq 0 \\ |a| & \text{otherwise} \end{cases} \qquad (17-3)$$

From the above it follows that

$$a = a^+ - a^- \qquad (17\text{-}4)$$

Based on the above relationship for month t the various cost functions are:

$C_1 W_t$ (regular payroll cost)

$C_{21}(W_t - W_{t-1})^+$ (hiring cost)

$C_{22}(W_t - W_{t-1})^-$ (layoff cost)

$C_3(kP_t - W_t)^+$ (overtime cost)

$C_{71} I_t^+$ (inventory cost)

$C_{72} I_t^-$ (shortage cost)

The Cost Model

The objective will be to minimize total cost over a period of n months. The cost model is formulated as,

Minimize $C = C(P_t; W_t, t = 1, 2, \ldots, n)$

$$= \sum_{t=1}^{n} [C_1 W_t + C_{21}(W_t - W_{t-1})^+ +$$

$$+ C_{22}(W_t - W_{t-1})^- + C_3(kP_t - W_t)^+ +$$

$$+ C_{71} I^+ + C_{72} I^-] \qquad (17\text{-}5)$$

subject to

$$P_t \geq 0 \qquad\qquad (17\text{-}6)$$

$$W_t \geq 0 \qquad\qquad (17\text{-}7)$$

$$I_t = I_{t-1} + P_t - S_t \quad , \ t = 1, \ 2, \ \ldots, n \qquad (17\text{-}8)$$

It is assumed that forecasts of S_t are available and the initial conditions (I_o, W_o) are known. Minimization of total cost over the n months is, of course with respect to the decision variables P_t and W_t.

In order to arrive at a linear cost function the following new set of variables are introduced. These are,

$$x_t = (W_t - W_{t-1})^+$$

$$y_t = (W_t - W_{t-1})^-$$

$$z_t = (kP_t - W_t)^+$$

$$w_t = (kP_t - W_t)^-$$

$$u_t = I_t^+$$

$$v_t = I_t^- \qquad , \ t = 1, 2, \ldots, n \qquad (17\text{-}9)$$

Based on the above notation P_t and W_t can be expressed as shown below. From equations (17-8) and (17-3) the following equality is derived,

$$P_t = I_t - I_{t-1} + S_t = (u_t - v_t) - (u_{t-1} - v_{t-1}) + S_t \qquad (17\text{-}10)$$

Similarly from equation (17-9) one can derive,

$$kP_t - W_t = z_t - w_t$$

$$W_t = kP_t - (z_t - w_t)$$

$$= k[(u_t - v_t) - (u_{t-1} - v_{t-1}) + S_t] - (z_t - w_t) \qquad (17\text{-}11)$$

The restrictions, equations (17-6) and (17-8), now take the form,

$$(u_t - v_t) - (u_{t-1} - v_{t-1}) + S_t \geq 0 \qquad (17\text{-}12)$$

and the restriction contained in equation (17-7) now reads,

$$(u_t - v_t) - (u_{t-1} - v_{t-1}) + S_t - (z_t - w_t)/k \geq 0$$

The definition of the variables x_t and y_t in equation (17-9) together with equation (17-4) mean that,

$$W_t - W_{t-1} = x_t - y_t$$

These restrictions may now be expressed in terms of the new variables. Using equation (17-11) the differences may be written in the form:

$$(W_t - W_{t-1})/k = (u_t - v_t) - 2(u_{t-1} - v_{t-1}) + (u_{t-2} - v_{t-2}) -$$

$$- (z_t - w_t)/k + (z_{t-1} - w_{t-1})/k + (S_t - S_{t-1})$$

Thus, the set of restrictions introduced by the definitions of the new variables is given by,

$$(u_t - v_t) - 2(u_{t-1} - v_{t-1}) + (u_{t-2} - v_{t-2}) - (z_t - w_t)/k +$$

$$+ (z_{t-1} - w_{t-1})/k - (x_t - y_t)/k = S_{t-1} - S_t \qquad (17\text{-}14)$$

Equation (17-2), (17-3) and (17-4) may be thought of as assumptions rather than restrictions since an optimum solution of a linear programming problem will automatically yield pairs of numbers (x_t, y_t) etc. with the property that either $x_t = 0$ or $y_t = 0$, etc. However, it is required that all these variables are non-negative:

$$x_t, \; y_t \geq 0$$

$$z_t, \; w_t \geq 0$$

$$u_t, \; v_t \geq 0 \qquad (17\text{-}15)$$

The cost function (17-5) can now be reformulated using equations (17-9) and (17-11) as follows:

$$C = \sum_{t=1}^{n} C_{21} x_t + C_{22} y_t + C_3 z_t + C_{71} u_t + C_{72} v_t +$$

$$+ C_1 k[(u_t - v_t) - (u_{t-1} - v_{t-1}) + S_t] - C_1 (z_t - w_t)$$

$$\text{or} \quad C = \sum_{t=1}^{n} [\quad C_{21} x_t +$$

$$+ \quad C_{22} y_t +$$

$$+ (C_3 - C_1) z_t +$$

$$+ \quad C_1 w_t +$$

$$+ \quad C_{71} u_t +$$

$$+ \quad C_{72} v_t] +$$

$$+ C_1 k[(u_n - v_n) - I_o + \sum_{t=1}^{n} S_t] \qquad (17\text{-}16)$$

In words the cost minimization can be stated as follows. Minimize the cost function (17-16) subject to the constraints of equations (17-12), (17-13), (17-14) and (17-15) for $t = 1, 2, \ldots, n$. Each month t of the planning interval contributes four variables and three restrictions. According to equations (17-12) and (17-13) the original decision variables P_t and W_t (production rate and workforce level for month t) are slack variables of the linear programming problem and can be obtained directly from the final simplex tableau.

AN EXAMPLE

Suppose that a firm engaged in the manufacture of widgets decides to apply the model discussed above to determine its optimal monthly workforce and production levels. It decides to use sales forecasts for three future months. The sales forecasts, current sales, current and previous month inventories and current overtime worked are given below.

$$S_0 = 24000 \qquad u_0 = 20000 \qquad z_0 = 0$$

$$S_1 = 25000 \qquad v_0 = 0 \qquad w_0 = 0$$

$$S_2 = 22000 \qquad u_{-1} = 18000$$

$$S_3 = 30000 \qquad v_{-1} = 0$$

The parameters in the cost model have the following values.

$C_1 = 4.00$ (dollars per hour)

$C_{21} = 500$ (dollars per worker hired)

$C_{22} = 300$ (dollars per worker laid off)

$C_3 = 6.00$ (dollars per overtime hour)

$C_{71} = 0.10$ (dollars per unit in inventory per month)

$C_{72} = 0.20$ (dollars per unit shortage per month)

$k = 2$ (number of hours required to produce one item of product)

$n = 3$ (number of months)

Using the model and the parameters listed above the problem is formulated as follows.

$$
\begin{aligned}
\text{Minimize } C = \;& 500\,x_1 + 300\,y_1 + 2\,z_1 + 4\,w_1 + 0.1\,u_1 + 0.2\,v_1 + \\
& + 500\,x_2 + 300\,y_2 + 2\,z_2 + 4\,w_2 + 0.1\,u_2 + 0.2\,v_2 + \\
& + 500\,x_3 + 300\,y_3 + 2\,z_3 + 4\,w_3 + 6.1\,u_3 - 5.9\,v_3
\end{aligned}
$$

subject to:

$$-u_1 + v_1 \leq 5000$$

$$u_1 - v_1 - u_2 + v_2 \leq 22000$$

$$u_2 - v_2 - u_3 + v_3 \leq 30000$$

$$- u_1 + v_1 + .5z_1 - .5w_1 \leq 5000$$

$$u_1 - v_1 - u_2 + v_2 + .5z_2 - .5w_2 \leq 22000$$

$$u_2 - v_2 - u_3 + v_3 + .5z_3 - .5w_3 \leq 3000$$

$$u_1 - v_1 - .5z_1 + .5w_1 - .5x_1 + .5y_1 = 21000$$

$$-2u_1 + 2v_1 + u_2 - v_2 + .5z_1 - .5w_1 - .5z_2 + .5w_2 - .5x_2 + .5y_2 = 17000$$

$$u_1 - v_2 - 2u_2 + 2v_2 + u_3 - v_3 + .5z_2 - .5w_2 - .5z_3 + .5w_3 - .5x_3 + .5y_3 = -8000$$

EXERCISES

1. Solve the problem formulated above for the work force and production levels for the next three months.

2. The sales forecast for a product produced by Continental Capacitors is as shown below.

Period	Units	Period	Units
1	1200	6	800
2	1100	7	800
3	1000	8	900
4	900	9	1000
5	850	10	1000

Regular production capacity is 1000 units per period and overtime capacity is 300 units per period. Capacity cannot be reduced below 400 units per period. Regular labor cost amounts to $750 per man per period. Overtime costs $25 per unit more than if produced on regular time. Storage costs $8 per unit per period and shortage costs $10 per unit per period. Hiring and layoff costs amount to $200 and $250 per man respectively. Solve the above problem and determine when and how much overtime must be worked by employees of Continental.

3. The union has indicated to Continental that it is willing to accept a nine hour day with the ninth hour being paid at 125 percent of regular pay. Presently all overtime is paid at 150 percent of regular pay. However, the union insists that Continental notify its labor force three periods ahead of time of changes and nine hour days must

last for a minimum of two periods. Should this union proposal be accepted or rejected by Continental?

4. What kind of proposal should Continental Capacitor present to the union assuming that it decides to reject the above proposal?

5. Suppose that Continental produces a product which cannot be held in inventory. In other words projected sales for a period must be produced during that period. To accomplish this the exact number of workers must be hired for that period. Alternatively overtime may be worked if the work force is insufficient or idle time will be incurred if excess workers are on the payroll. Redesign the model if necessary and solve Continental's work force planning problem.

6. Develop a modified version of the linear programming model to the work force and inventory planning cost function developed in the previous chapter and of the form,

$$E(\cdot) = \sum_{t=1}^{T} [C_1 P_t / C_4$$
$$+ C_2 (W_t - W_{t-1})^2$$
$$+ C_3 (P_t - C_4 W_t)^2]$$

subject to $S_t = P_t$, for $t = 1, 2, \ldots, T$

7. Apply the modified linear programming model developed in exercise 6 to the problem listed in Table 16-1 of the previous chapter. The

parameters are $C_1 = 500$, $C_2 = 5$, $C_3 = 50$ and $C_4 = 1.428$. Compare your results with the results in Table 16-2.

CHAPTER 18.--WORK FORCE AND PRODUCTION PLANNING THROUGH
INVENTORY PROJECTION BY INPUT-OUTPUT ANALYSIS

In the previous two chapters models were presented
which determined the optimal production, work force and
inventory levels of a manufacturing firm under conditions
of varying demand over time, especially in terms of severe
seasonal variations in demand during the year. The model
discussed in this chapter determines the levels of investment in the various categories of inventory under conditions of non-cyclical demand over the planning period.
The model was proposed by Smith[1] and uses the technique
of input-output analysis originally proposed by Leontief.[2]

BACKGROUND AND MODEL ENVIRONMENT

The input-output technique was originally used for
determining flows between industries in the national
economy. In this chapter the approach is based on the use
of historical or pre-determined relationships among the
flows and accumulations of value throughout the inventory
system modified by the estimated effects of any changes
anticipated in the operation of the system.

[1]S.B. Smith, "An Input-Output Model for Production and
Inventory Planning," Journal of Industrial Engineering,
January-February 1965, pp. 64-69.

[2]Wassily Leontief, The Structure of the American
Economy, 1919-1939, Second Edition, New York: Oxford
University Press, 1951, pp. 143-46.

Model Environment

The inventory planning model described in this chapter is based on a machinery manufacturer producing and selling four related but separate categories of products. These are machinery, subassemblies and two categories of parts. The subassemblies are made up of individual parts and the machinery is made up of subassemblies and individual parts. The latter two are obviously not completely mutually exclusive. Sale of parts and subassemblies could be considered to be in response to replacement components for the previously-sold machinery. However, it is conceivable that individual parts and subassemblies are being used in the manufacture of machinery by another manufacturer. One of the assigned problems in the exercises section assumes this explicitly. Hence, one can see that the firm for which the model has been developed is quite typical and as such the model is quite general and has many potential applications.

It will be assumed that the system is in equilibrium on a monthly basis. That is the sum of all inputs during a given month is equal to cost of sales for that month and inputs to each inventory category equal the outputs of that inventory category.

Background

The production plan in a manufacturing firm is based on the sales forecast plus or minus the projected change in the level of inventory. With large inventory invest-ments due to long manufacturing lead times, planned in-ventory changes can have a substantial effect on the level of production.

Manufacturing firms producing only a few items can prepare an inventory plan by summing the required inventory for each item through all stages of manufacturing. Firms producing a large number of items commonly use aggregate production and inventory relationships. Only in cases where the product mix tends to vary widely is an aggregate approach undesirable.

The model in this chapter presents a method of planning which uses aggregate production and inventory relationships together with sales forecasts and is based on an input-output model of the firm's production and inventory system.

MODEL DEVELOPMENT

The production-inventory system used for illustrating the model is exhibited in Figure 18-1. Note that there are two categories of inputs into the system. The first category consists of external resources or purchases which

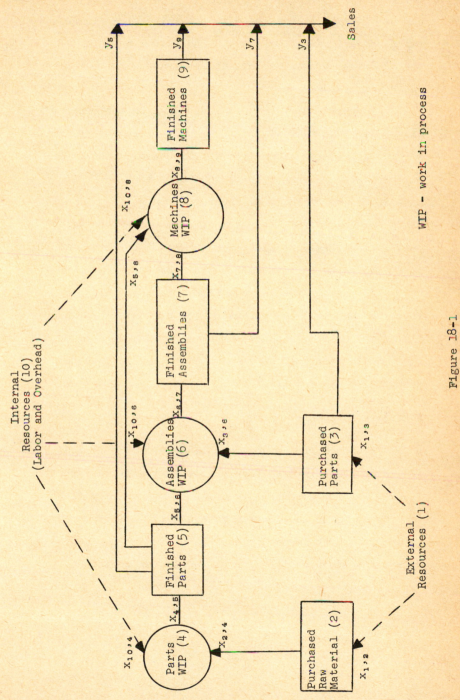

Figure 18-1

Flow Diagram of a Production Inventory System

WIP - work in process

can be directly associated with the inventory. These
purchases are raw materials and individual parts. The
second category consists of internal resources such as
labor and overhead.

In this model overhead is applied on the basis of
direct labor worked and both labor and overhead can therefore be considered as one input. If overhead is not applied
on the basis of direct labor, the model must be modified
with labor and overhead as separate inputs.

Outputs from the system consist of manufactured
parts, assemblies, machines and purchased parts. Note that
purchased parts is both an input and an output. Hence the
model is also applicable to a firm which partially manufactures and partially purchases its products.

The unit of measurement for input as well as output
is in dollars of cost per month. Outputs are thus not
measured in sales dollars but in cost of sales dollars.

Inventory Categories

Although Figure 18-1 has been labeled a flow diagram
of a production-inventory system it is really an inventory
diagram for the purpose of representing inventory flow.
The inventory in the diagram consists of eight inventory
categories. Categories in squares or rectangles represent
stores inventory and categories in circles represent

work-in-process (WIP) inventories. Moving from left to
right it may be observed that the raw material upon delivery
is stored before it proceeds to be processed into parts
and thus becomes WIP inventory. Upon completion of the
parts manufacturing process the finished parts are stored.
The finished parts in storage then may proceed into three
possible directions. They are sold, enter the machine
WIP inventory or enter the assemblies WIP inventory. At
this point purchased parts also move from purchased parts
inventory to assemblies WIP inventory. Upon completion
the finished assemblies are stored in finished assemblies
inventory from where they are either sold or enter machines
WIP inventory. Finished machines, the last inventory station, of course proceed to the customer. One can thus see
that eight inventory stations may be identified in the
production-inventory system diagram.

Inventory Valuation

A projection of the inventory level will be made
at constant dollars for each of the eight inventory categories. This projection represents physical changes of
inventory and therefore may be used directly for calculating
direct man hours which should be added or deducted from
hours required to meet sales in planning personnel requirements. As responsibility for physical changes in inventory

level rests with managers it also provides a control on
their decisions related to inventory levels.

For projecting inventory flows at constant dollars,
three sets of data are required. These are, 1) a forecast
of cost of sales at constant dollars for each category of
inventory; 2) the input fractions, also called production
technology, expressed in terms of the dollar inputs of
labor, overhead, and material as proportions of the total
input to each category of inventory; and 3) the prescribed
inventory policies.

As was stated previously the total system and subsystems are assumed to be in equilibrium on a monthly
basis. For each month the inputs to each inventory category must thus be determined. These inputs consist of
labor, overhead and material or previously-manufactured
parts or assemblies.

Mathematical Relationships

Before presenting the mathematical relationships the
following symbols will be defined. Let

y_i = sales at cost per month from inventory i

x_{ij} = flow of value at cost per month from inventory
i to inventory j

a_{ij} = input from inventory i to inventory j expressed
as a proportion of the total inputs to inventory j

X_i = total output at cost per month from inventory i

m = total number of inventory categories plus the two inputs from internal (labor and overhead) and external (purchases) sources

The total output of each inventory category is the sum of its sales (in cost of sales terms) and inputs to other inventory categories. This may be stated as,

$$X_1 = y_i + \sum_{j=1}^{m} x_{ij} \qquad (18\text{-}1)$$

But under the equilibrium assumptions and from the definition of a_{ij},

$$x_{ij} = a_{ij} X_j \qquad (18\text{-}2)$$

Combining equations (18-1) and (18-2) results in,

$$X_i = y_i + \sum_{j=1}^{m} a_{ij} X_j . \qquad (18\text{-}3)$$

Utilizing equation (18-3), equations for each of the X_i's for the inventory system portrayed in Figure 18-1 can be developed as follows,

$$X_1 = a_{1,2} X_2 + a_{1,3} X_3$$

$$X_2 = a_{2,4} X_4$$

$$X_3 = a_{3,6} X_6 + y_3$$

$$X_4 = a_{4,5} \ X_5$$

$$X_5 = a_{5,6} \ X_6 + a_{5,8} \ X_8 + y_5$$

$$X_6 = a_{6,7} \ X_7$$

$$X_7 = a_{7,8} \ X_8 + y_7$$

$$X_8 = a_{8,9} \ X_9$$

$$X_9 = y_9$$

$$X_{10} = a_{10,4} \ X_4 + a_{10,6} \ X_6 + a_{10,8} \ X_8 \qquad (18\text{-}4)$$

The relationships in (18-4) provide ten equations in ten unknowns. Solving for the X_i's results in:

$$X_1 = a_{2,4}[y_5 + a_{5,6}(y_7 + a_{7,8}y_9) + a_{5,8}y_9] + y_3 + a_{3,6}(y_7 + a_{7,8}y_9)$$

$$X_2 = a_{2,4}[y_5 + a_{5,6}(y_7 + a_{7,8}y_9) + a_{5,8}y_9]$$

$$X_3 = y_3 + a_{3,6}(y_7 + a_{7,8}y_9)$$

$$X_4 = y_5 + a_{5,6}(y_7 + a_{7,8}y_9) + a_{5,8}y_9$$

$$X_5 = y_5 + a_{5,6}(y_7 + a_{7,8}y_9) + a_{5,8}y_9 \qquad (18\text{-}5)$$

$$X_6 = y_7 + a_{7,8}y_9$$

$$X_7 = y_7 + a_{7,8}y_9$$

$$X_8 = y_9$$

$$X_9 = y_9$$

$$X_{10} = a_{10,4}y_5+(a_{10,8}+a_{5,8}a_{10,4})y_9+(a_{10,6}+a_{5,6}a_{10,4})(y_7+a_{7,8}y_9)$$

Utilizing equation (18-2) the value of each resource flow in the production-inventory system may be calculated.

DETERMINATION OF INVENTORY VALUES

The resource flows determined in the previous section will now be used to determine the inventory levels for each inventory category. However, to determine inventory levels the firm's inventory policies and some knowledge of the production process must be available.

Stores Inventory

Stores inventories (the rectangles in Figure 18-1) will be analyzed first. Inventories in stores can be viewed as consisting of a turnover portion and a buffer or reserve portion. The latter would of course be determined by the firm's inventory policy and will be identified by the symbol R where R is measured in terms of months' supply. The turnover portion is analogous to the economic lot size or order size. If the order size is Q in terms of months' supply, then the average inventory in stores will be Q/2 for turnover purposes. Total average inventory will then be Q/2 +R in terms of months' supply.

Assuming that for each inventory category the turn-over and buffer inventory in terms of months' supply can be determined, then the total expected value of inventory, I_j, for inventory category j will be

$$I_j = (Q_j/2 + R_j) \, X_j \qquad (18\text{-}6)$$

As the system is supposed to be in equilibrium, the total output from inventory j per month is equal to the sum of the inputs to inventory j per month. Therefore, equation (18-6) can be restated as,

$$I_j = (Q_j/2 + R) \sum_{i=1}^{m} x_{ij} \qquad (18\text{-}7)$$

Finished Machines Inventory

For some inventory categories equations (18-6) and (18-7) may not adequately represent the inventory levels. For instance the inventory level of finished machines is determined by the number of machines that are kept in inventory, N, to support the level of forecasted sales. If n is the average number of machines to be produced per month, the expected value of finished machines in stores can be represented as follows,

$$I_9 = Nx_{8,9}/n \qquad (18\text{-}8)$$

The value $x_{8,9}/n$ in equation (18-8) is the average value of a machine at cost.

Work in Process Inventory

The value of inventories of work in process is determined by the speed at which value is added to an order.
Typically raw material is added as soon as the production
order is issued. However, this does not always hold.
Labor and overhead on the other hand are added as work
progresses on an order. If it is assumed that material is
added at the start of an order and labor and overhead are
added at a constant rate during the duration of the lead
time the condition as portrayed in Figure 18-2 prevails.
If in addition material cost is equal to labor and overhead
it can be concluded that the average percent of value added
during the entire lead time is 75 percent.

Average Percent of Value Added

Average percent of value added is a useful concept
and is easy to determine if the conditions are similar
to the above example. However, the rate of material,
labor and overhead addition is usually not linear or instantaneous and other techniques are necessary to estimate
the average percent of value added. One way to accomplish
this is by the following model. Let average percent value
added be p, then

$$p = \int_{0}^{T} 100V(t)\ dt / [TV(T)] \qquad (18\text{-}9)$$

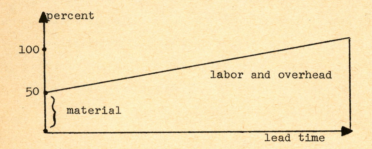

Figure 18-2

Illustration of Value Added to a Product

where T is the lead time, $V(t)$ is the value added to the
order by time t and $V(T)$ is total value added to the order.

The reader may respond that equation (18-9) does
not help him much because he does not know the function
$V(t)$. This is true and in practice it is up to the individual doing the analysis to approximate $V(t)$ with a
straight line if it is a discontinuous function or alternatively to estimate the average percent value added
directly from the information of the production process
available to him.

From the above analysis and with reference to the
flows in Figure 18-1 levels of inventory can be estimated
for all inventory categories. The formulas to accomplish
this are given in Table 18-1.

Table 18-1

Formulas for Estimating Inventory Levels

Inventory	Symbol	Formula
Raw Material	I_2	$x_{1,2}(R_2+Q_2/2)$
Purchased Parts	I_3	$x_{1,3}(R_3+Q_3/2)$
Parts Work in Process	I_4	$T_4 p_4 (x_{2,4}+x_{10,4})$
Finished Parts	I_5	$x_{4,5}(R_5+Q_5/2)$
Assemblies Work in Process	I_6	$T_6 p_6 (x_{5,6}+x_{3,6}+x_{10,6})$
Finished Assemblies	I_7	$x_{6,7}(R_7+Q_7/2)$
Machines in Process	I_8	$T_8 p_8 (x_{5,8}+x_{7,8}+x_{10,8})$
Finished Machines	I_9	$N x_{8,9}/n$

EXAMPLE

To illustrate the model discussed in this chapter an example will be worked out in detail below. The sales at cost in dollars per month for the example are $3000 for purchased parts, $5000 for finished parts, $1000 for finished assemblies, and $20,000 for finished machines.

The input ratios are listed below.

$a_{1,2} = 1.0$	$a_{3,6} = 0.2$	$a_{7,8} = 0.6$
$a_{1,3} = 1.0$	$a_{5,6} = 0.5$	$a_{5,8} = 0.2$
$a_{2,4} = 0.7$	$a_{10,6} = 0.3$	$a_{10,8} = 0.2$
$a_{10,4} = 0.3$	$a_{6,7} = 1.0$	$a_{8,9} = 1.0$
$a_{4,5} = 1.0$		

Using equations (18-5) the total outputs at cost in dollars per month, X_i for inventory i, i=1,2,...,10, can be found. The respective outputs are shown below.

$X_1 = \$16450$	$X_5 = \$15500$	$X_8 = \$20000$
$X_2 = \$10850$	$X_6 = \$13000$	$X_9 = \$20000$
$X_3 = \$5600$	$X_7 = \$13000$	$X_{10} = \$12550$
$X_4 = \$15500$		

There are several checks that can be made on the data and calculations made thus far. Note that the input ratios a_{ij} must satisfy the constraint,

$$\sum_{i=1}^{m} a_{ij} = 1 \quad \text{for } j = 1, 2, \ldots, m.$$

Also since a closed system is being analyzed the sum of all inputs in dollars per month must equal the sum of all outputs in dollars per month. The sum of all inputs consists of external resources such as purchases, X_1, and internal resources such as labor and overhead, X_{10}. The sum of all outputs consist of purchased parts sales, y_3, finished parts sales, y_5, finished assemblies sales, y_7 and finished machines sales, y_9. Applying this check it may be observed that $X_1 + X_{10} = \$29000$ and $y_3 + y_5 + y_7 + y_9 = \$29000$.

To find the flow of value at cost in dollars per month, x_{ij}, from inventory i to inventory j formula (18-2) is applied and the results are shown below.

$x_{1,2} = \$10850$	$x_{5,6} = \$6500$	$x_{8,9} = \$20000$
$x_{1,3} = \$5600$	$x_{5,8} = \$4000$	$x_{10,4} = \$4650$
$x_{2,4} = \$10850$	$x_{6,7} = \$13000$	$x_{10,6} = \$3900$
$x_{3,6} = \$2600$	$x_{7,8} = \$12000$	$x_{10,8} = \$4000$
$x_{4,5} = \$15500$		

EXERCISES

1. Solve the machinery plant input-output problem used in this chapter with the required monthly outputs at cost in dollars and the parameters as shown below.

$y_3 = \$5000$; $y_5 = \$2000$; $y_7 = \$25000$; $y_9 = \$10000$

$a_{1,2} = 1.0$	$a_{3,6} = 0.4$	$a_{7,8} = 0.5$
$a_{1,3} = 1.0$	$a_{5,6} = 0.4$	$a_{5,8} = 0.4$
$a_{2,4} = 0.2$	$a_{10,6} = 0.2$	$a_{10,8} = 0.1$
$a_{10,4} = 0.8$	$a_{6,7} = 1.0$	$a_{8,9} = 1.0$
$a_{4,5} = 1.0$		

2. Based on the flow diagram for a machinery manufacturing plant shown in Figure 18-3 and the monthly output data at cost in dollars and parameters shown below solve for total output at cost per month for each inventory and also solve for the value flows between inventories.

$y_3 = \$2000$	$y_7 = \$5000$	$y_{11} = \$25000$
$y_4 = \$1000$	$y_9 = \$4000$	

$a_{1,2} = 1.0$	$a_{3,8} = 0.4$	$a_{7,10} = 0.1$
$a_{1,3} = 1.0$	$a_{4,8} = 0.1$	$a_{9,10} = 0.4$
$a_{1,4} = 1.0$	$a_{7,8} = 0.2$	$a_{12,10} = 0.3$
$a_{1,5} = 1.0$	$a_{12,8} = 0.3$	$a_{6,7} = 1.0$
$a_{2,6} = 0.6$	$a_{4,10} = 0.1$	$a_{8,9} = 1.0$
$a_{12,6} = 0.4$	$a_{5,10} = 0.1$	$a_{10,11} = 1.0$

3. Draw an input-output flow diagram for an automobile manufacturer who sells five groups of products. The five groups of products with their respective sales at cost in dollars per month are shown below.

449

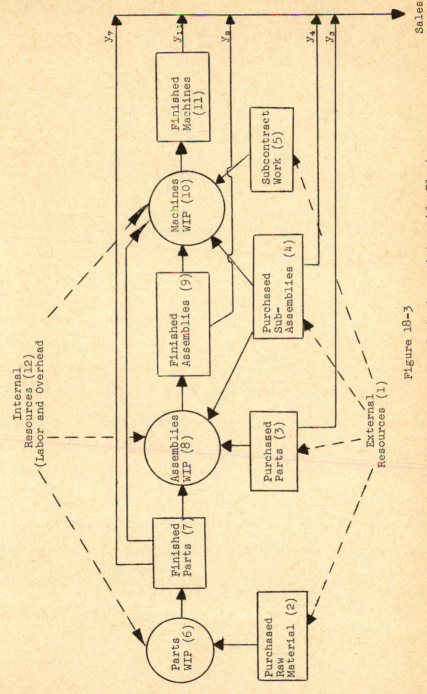

Figure 18-3

Flow Diagram of Machine Manufacturing and Assembly Plant

Finished engines   - $100,000

Finished transmissions - $ 50,000

Finished automobiles   - $600,000

Parts   - $ 40,000

Assemblies   - $ 80,000

The inputs into the manufacturing plant consist of external resources such as raw material, parts and sub-assemblies, and of internal resources such as labor and overhead.

4. Assign reasonable values to the inventory proportions, the a_{ij}'s in problem 3. Then solve for the total output at cost per month for all inventories, the X_i's and for all inventory flows at cost per month, the x_{ij}'s. What is the total monthly cost for labor and overhead?

5. National Bearing Corporation produces ball and roller bearings for a variety of customers. Because of inadequate production capacity National purchases 1/3 of its steel ball requirements and 1/2 of its steel roller requirements from vendors. Total annual sales at cost amount to $4.5 million of ball bearings and $9.7 million of roller bearings. Labor input amounts to $3.5 million annually. Material input excluding steel balls and rollers amounts to $6.4 million annually and annual overhead amounts to an additional $2.0 million. Draw the input-output flow diagram for National Bearing Corporation.

6. Suppose sales of roller bearings would double for National Bearing Corporation from $9.7 million to $19.4 million annually while ball bearing sales remained the same. How would this affect material input, steel roller purchases, overhead and labor? State your assumptions.

7. Premier Electric assembles television sets, transistor radios and stereo components. It also produces 10 percent of the components for its various products; the remainder is purchased from vendors. Premier's products are guaranteed for parts only and it maintains an extensive service facility to service inoperative products. Draw the input-output diagram for Premier Electric.

8. Suppose monthly sales at cost of Premier Electric are as shown below.

Television sets	$1,050,000
Transistor radios	400,000
Stereo components	975,000
Repair services	75,000

Solve for the x_{ij}'s. State your assumptions.

9. Suburban Products produces burglar alarm devices for household and commercial uses. Suburban is essentially an assembler of various components into kits. A complete kit consists of a sufficient number of components and assemblies to service a house or a business. The sizes of the kits vary depending on the size of the house or the business being serviced. The components purchased consist of four

basic categories. These are wiring, hardware, activators and alarm units. Draw the input-output flow diagram for Suburban.

10. Monthly sales at cost of Suburban Products amount to $1,900,000. Solve for the x_{ij}'s assuming your own values of a_{ij}'s. What is the total monthly cost for labor and overhead under your assumptions?

11. Suppose a fraction 1-k, where k < 1, of inputs to each rectangle or circle in the flow diagram, Figure 18-3, were lost due to spoilage, scrap or rejection. Can the input-output analysis still be used? If it can still be used, show how the mathematical relationships will change.

12. Solve exercise 3 assuming that the rejection fraction 1-k equals .02, or k = .98.

CORRUGATOR SIMULATION PROGRAM
(referenced in Chapter 2)

```
C       CORRUGATOR ORDER GENERATOR AND SCHEDULER
        DIMENSION Y(200),Q(200),QA(200),QB(200),QC(200),W(200),WA(200),
       2WB(200),WC(200),AL(200),ALB(200),ALC(200),TWID(200),TLEN(200),
       3ALA(200),TLENG(200),SIGAQ(200),SIGAW(200),SIGAL(200),AVAQ(200),
       4AVAW(200),AVAL(200),RHO(200),WW(200),ALL(200)
        READ(5,3) NA,NB,NC,ND,JDOALL,WYD,WWYD
      3 FORMAT(5I10,2F10.2)
        READ(5,4)(SIGAQ(JL),SIGAW(JL),SIGAL(JL),AVAQ(JL),AVAW(JL),
       2AVAL(JL),RHO(JL),JL=1,25)
      4 FORMAT(7F10.3,2X)
        WRITE(6,630)NA,NB,NC,ND,JDOALL,WYD,WWYD
    630 FORMAT(46H NUMBER OF ORDERS PER FLUTE A B C DW    CYCLES//(5I10,
       22F10.2))
        WRITE(6,339)(SIGAQ(I),SIGAW(I),SIGAL(I),AVAQ(I),AVAW(I),AVAL(I),
       2RHO(I),I=1,25)
    339 FORMAT(7F10.3)
        A=WYD
        GTPWCH=0.0
        GTSUP=0.0
        GTTIM=0.0
        GTLENG=0.0
        GTAREA=0.0
        GTPCSP=0.0
        GTSCRP=0.0
        LATOT=0
        LBTOT=0
        JALLAG=1
      5 KSTATM=0
        N=0
        JABCAG=1
     10 DO 20 I=1,NA
        Y(I)=.5*(FLRAN(X)+1.)
        IF(Y(I)-.24)15,15,20
     15 N=N+1
     20 CONTINUE
        FLUT=1.
        JL=1
        GO TO 888
    801 WRITE(6,25)
     25 FORMAT(18X,35H SCHEDULE FOR FLUTE A AND GRADE KXK)
        GO TO 425
    130 DO 145 I=1,NB
        Y(I)=.5*(FLRAN(X)+1.)
        IF(Y(I)-.11)140,140,145
    140 N=N+1
    145 CONTINUE
        FLUT=1.
        JL=2
        GO TO 888
    802 WRITE(6,150)
    150 FORMAT(18X,35H SCHEDULE FOR FLUTE B AND GRADE KXK)
        GO TO 425
    189 DO 192 I=1,NB
        Y(I)=.5*(FLRAN(X)+1.)
        IF(Y(I)-.30)192,192,190
    190 IF(Y(I)-.37)191,191,192
    191 N=N+1
    192 CONTINUE
        FLUT=1.
        JL=3
        GO TO 888
    803 WRITE(6,193)
    193 FORMAT( 18X,35H SCHEDULE FOR FLUTE B AND GRADE KXM)
        GO TO 425
    180 DO 187 I=1,NB
        Y(I)=.5*(FLRAN(X)+1.)
        IF(Y(I)-.07)187,187,185
    185 IF(Y(I)-.14)186,186,187
    186 N=N+1
    187 CONTINUE
        FLUT=1.
        JL=4
        GO TO 888
    804 WRITE(6,188)
    188 FORMAT(18X,35H SCHEDULE FOR FLUTE B AND GRADE NXN)
        GO TO 425
    255 DO 270 I=1,NC
        Y(I)=.5*(FLRAN(X)+1.)
        IF(Y(I)-.24)270,270,260
```

```
260 IF(Y(I)-.287)265,265,270
265 N=N+1
270 CONTINUE
    FLUT=1.
    JL=5
    GO TO 888
805 WRITE(6,275)
275 FORMAT(18X,37H SCHEDULE FOR FLUTE C AND GRADE CWKXK)
    GO TO 425
208 DO 215 I=1,NC
    Y(I)=.5*(FLRAN(X)+1.)
    IF(Y(I)-.17)210,210,215
210 N=N+1
215 CONTINUE
    FLUT=1.
    JL=6
    GO TO 888
806 WRITE(6,220)
220 FORMAT(18X,35H SCHEDULE FOR FLUTE C AND GRADE KXK)
    GO TO 425
280 DO 295 I=1,NC
    Y(I)=.5*(FLRAN(X)+1.)
    IF(Y(I)-.31)295,295,285
285 IF(Y(I)-.327)290,290,295
290 N=N+1
295 CONTINUE
    FLUT=1.
    JL=7
    GO TO 888
807 WRITE(6,300)
300 FORMAT(18X,35H SCHEDULA FOR FLUTE C AND GRADE KXM)
    GO TO 425
305 DO 320 I=1,NC
    Y(I)=.5*(FLRAN(X)+1.)
    IF(Y(I)-.37)320,320,310
310 IF(Y(I)-.42)315,315,320
315 N=N+1
320 CONTINUE
    FLUT=2.
    JL=8
    GO TO 888
808 WRITE(6,325)
325 FORMAT(18X,35H SCHEDULE FOR FLUTE C AND GRADE JXJ)
    GO TO 425
350 JABCAG=JABCAG+1
    IF(JABCAG-3)10,10,80
 80 DO 95 I=1,NA
    Y(I)=.5*(FLRAN(X)+1.)
    IF(Y(I)-.02)90,95,95
 90 N=N+1
 95 CONTINUE
    FLUT=1.
    JL=9
    GO TO 888
809 WRITE(6,100)
100 FORMAT(18X,35H SCHEDULE FOR FLUTE A AND GRADE MXM)
    GO TO 425
 55 DO 70 I=1,NA
    Y(I)=.5*(FLRAN(X)+1.)
    IF(Y(I)-.29)70,70,60
 60 IF(Y(I)-.35)65,65,70
 65 N=N+1
 70 CONTINUE
    FLUT=2.
    JL=10
    GO TO 888
810 WRITE(6,75)
 75 FORMAT(18X,35H SCHEDULE FOR FLUTE A AND GRADE JXJ)
    GO TO 425
105 DO 110 I=1,NA
    Y(I)=.5*(FLRAN(X)+1.)
    IF(Y(I)-.06)110,110,106
106 IF(Y(I)-.10)107,107,110
107 N=N+1
110 CONTINUE
    FLUT=2.
    JL=11
    GO TO 888
811 WRITE(6,115)
```

```
115 FORMAT(18X,35H SCHEDULE FOR FLUTE A AND GRADE JXK)
    GO TO 425
120 DO 125 I=1,NA
    Y(I)=.5*(FLRAN(X)+1.)
    IF(Y(I)-.10)125,125,122
122 IF(Y(I)-.14)123,123,125
123 N=N+1
125 CONTINUE
    FLUT=1.
    JL=12
    GO TO 888
812 WRITE(6,126)
126 FORMAT(18X,37H SCHEDULE FOR FLUTE A AND GRADE CWKXK)
    GO TO 425
155 DO 170 I=1,NA
    Y(I)=.5*(FLRAN(X)+1.)
    IF(Y(I)-.21)170,170,160
160 IF(Y(I)-.24)165,165,170
165 N=N+1
170 CONTINUE
    FLUT=1.
    JL=13
    GO TO 888
813 WRITE(6,175)
175 FORMAT(18X,35H SCHEDULE FOR FLUTE A AND GRADE KXM)
    GO TO 425
 30 DO 45 I=1,NA
    Y(I)=.5*(FLRAN(X)+1.)
    IF(Y(I)-.15)45,45,35
 35 IF(Y(I)-.24)40,40,45
 40 N=N+1
 45 CONTINUE
    FLUT=1.
    JL=14
    GO TO 888
814 WRITE(6,50)
 50 FORMAT(18X,35H SCHEDULE FOR FLUTE A AND GRADE NXN)
    GO TO 425
199 DO 202 I=1,NB
    Y(I)=.5*(FLRAN(X)+1.)
    IF(Y(I)-.45)202,202,200
200 IF(Y(I)-.51)201,201,202
201 N=N+1
202 CONTINUE
    FLUT=1.
    JL=15
    GO TO 888
815 WRITE(6,203)
203 FORMAT(18X,35H SCHEDULE FOR FLUTE B AND GRADE MXN)
    GO TO 425
204 DO 206 I=1,NB
    Y(I)=.5*(FLRAN(X)+1.)
    IF(Y(I)-.05)205,205,206
205 N=N+1
206 CONTINUE
    FLUT=1.
    JL=16
    GO TO 888
816 WRITE(6,207)
207 FORMAT(18X,35H SCHEDULE FOR FLUTE B AND GRADE MXM)
    GO TO 425
330 DO 340 I=1,NB
    Y(I)=.5*(FLRAN(X)+1.)
    IF(Y(I)-.02)340,340,333
333 IF(Y(I)-.05)335,335,340
335 N=N+1
340 CONTINUE
    FLUT=1.
    JL=17
    GO TO 888
817 WRITE(6,345)
345 FORMAT(18X,37H SCHEDULA FOR FLUTE B AND GRADE CWKXM)
    GO TO 425
405 DO 410 I=1,NB
    Y(I)=.5*(FLRAN(X)+1.)
    IF(Y(I)-.04)410,410,406
406 IF(Y(I)-.07)407,407,410
407 N=N+1
410 CONTINUE
```

```
          FLUT=2.
          JL=18
          GO TO 888
 818      WRITE(6,411)
 411      FORMAT(18X,37H SCHEDULE FOR FLUTE B AND GRADE CWJXJ)
          GO TO 425
 194      DO 197 I=1,NB
          Y(I)=.5*(FLRAN(X)+1.)
          IF(Y(I)-.42)197,197,195
 195      IF(Y(I)-.50)196,196,197
 196      N=N+1
 197      CONTINUE
          FLUT=2.
          JL=19
          GO TO 888
 819      WRITE(6,198)
 198      FORMAT(18X,35H SCHEDULE FOR FLUTE B AND GRADE JXJ)
          GO TO 425
 230      DO 240 I=1,NC
          Y(I)=.5*(FLRAN(X)+1.)
          IF(Y(I)-.22)240,240,233
 233      IF(Y(I)-.30)235,235,240
 235      N=N+1
 240      CONTINUE
          FLUT=1.
          JL=20
          GO TO 888
 820      WRITE(6,245)
 245      FORMAT(18X,35H SCHEDULE FOR FLUTE C AND GRADE NXN)
          GO TO 425
 412      DO 414 I=1,NC
          Y(I)=.5*(FLRAN(X)+1.)
          IF(Y(I)-.04)413,413,414
 413      N=N+1
 414      CONTINUE
          FLUT=2.
          JL=21
          GO TO 888
 821      WRITE(6,415)
 415      FORMAT(18X,35H SCHEDULE FOR FLUTE C AND GRADE KXJ)
          GO TO 425
 612      DO 614 I=1,NC
          Y(I)=.5*(FLRAN(X)+1.)
          IF(Y(I)-.03)614,614,611
 611      IF(Y(I)-.05)613,613,614
 613      N=N+1
 614      CONTINUE
          FLUT=2.
          JL=22
          GO TO 888
 822      WRITE(6,615)
 615      FORMAT(18X,37H SCHEDULE FOR FLUTE C AND GRADE CWJXJ)
          GO TO 425
 512      DO 514 I=1,NC
          Y(I)=.5*(FLRAN(X)+1.)
          IF(Y(I)-.01)514,514,511
 511      IF(Y(I)-.02)513,513,514
 513      N=N+1
 514      CONTINUE
          FLUT=2.
          JL=23
          GO TO 888
 823      WRITE(6,515)
 515      FORMAT(18X,35H SCHEDULE FOR FLUTE C AND GRADE HXH)
          GO TO 425
 355      DO 370 I=1,ND
          Y(I)=.5*(FLRAN(X)+1.)
          IF(Y(I)-.55)365,365,370
 365      N=N+1
 370      CONTINUE
          FLUT=3.
          JL=24
          GO TO 888
 824      WRITE(6,375)
 375      FORMAT(18X,38H SCHEDULE FOR FLUTE AB AND GRADE MXNXM)
          GO TO 425
 380      DO 395 I=1,ND
          Y(I)=.5*(FLRAN(X)+1.)
          IF(Y(I)-.55)395,395,390
```

```
390 N=N+1
395 CONTINUE
    FLUT=3.
    JL=25
    GO TO 888
825 WRITE(6,400)
400 FORMAT(18X,38H SCHEDULE FOR FLUTE AB AND GRADE MXMXM)
    GO TO 425
888 IF(N)890,890,889
890 KSTATM=KSTATM+1
    GO TO 891
889 CALL GEN(SIGAQ(JL),SIGAW(JL),SIGAL(JL),AVAQ(JL),AVAW(JL),
   2AVAL(JL),RHO(JL),Q,W,AL,WW,ALL,N,M,WWYD)
     CALL SCHED(A,Q,QA,QB,QC,W,WA,WB,WC,AL,ALA,ALB,ALC,TWID,N,K,
   2TLENG,TOTSUT,WW,ALL,M,LISTB,WYD,WWYD,FLUT)
     GO TO (801,802,803,804,805,806,807,808,809,810,811,812,813,814,
   2815,816,817,818,819,820,821,822,823,824,825),JL
425 AREA=0.0
    SCRAP=0.0
    TOTLEN=0.0
    LISTA=N
    DO 426 I=1,K
    TOTLEN=TOTLEN+TLENG(I)
    AREA=AREA+TLENG(I)*TWID(I)/12.
426 SCRAP=SCRAP+TLENG(I)*TWID(I)/12.-(QA(I)*WA(I)*ALA(I)+QB(I)*WB(I)*
   2ALB(I)+QC(I)*WC(I)*ALC(I))/144.
    RLCTMS=12.
    RLCTMD=15.
    SWFPMK=315.
    SWFPMJ=185.
    DWFPM=105.
    IF(FLUT-2.)424,427,428
424 TOTTIM=TOTLEN/SWFPMK
    DO 580 I=1,K
    IF(TLENG(I)-SWFPMK*RLCTMS)575,580,580
575 TOTSUT=TOTSUT+(SWFPMK*RLCTMS-TLENG(I))/SWFPMK
580 CONTINUE
    GO TO 429
427 TOTTIM=TOTLEN/SWFPMJ
    DO 780 I=1,K
    IF(TLENG(I)-SWFPMJ*RLCTMS)775,780,780
775 TOTSUT=TOTSUT+(SWFPMJ*RLCTMS-TLENG(I))/SWFPMJ
780 CONTINUE
    GO TO 429
428 TOTTIM=TOTLEN/DWFPM
    DO 880 I=1,K
    IF(TLENG(I)-DWFPM*RLCTMD)875,880,880
875 TOTSUT=TOTSUT+(DWFPM*RLCTMD-TLENG(I))/DWFPM
880 CONTINUE
429 AVTRIM=AREA*12./TOTLEN
    PWIDCH=1.0
    IF(K-1)481,481,431
431 KK=K-1
    DO 480 I=1,KK
    IF(TWID(I)-TWID(I+1))480,480,475
475 PWIDCH=PWIDCH+1.
480 CONTINUE
481 TOTSUT=TOTSUT+10.
    YLUDOX=.5*(FLRAN(X)+1.)
    IF(YLUDOX-.25)881,881,882
881 TOTSUT=TOTSUT+15.
882 TOTTIM=TOTTIM+TOTSUT
    RWID=AREA/TOTLEN
    RLSCRP=TOTLEN/5000.*RWID*10.+PWIDCH*RWID*25.
    ENDSP=TOTLEN/5.*RWID*.055
    TOTLEN=TOTLEN+(RLSCRP+ENDSP)/RWID
    AREA=AREA+RLSCRP+ENDSP
    SCRAP=SCRAP+RLSCRP+ENDSP
    PCSCRP=SCRAP*100./AREA
    WRITE(6,430)(QA(I),WA(I),ALA(I),QB(I),WB(I),ALB(I),QC(I),WC(I),
   2ALC(I),TLENG(I),TWID(I),I=1,K)
430 FORMAT(72H  QUANT WIDTH LENGTH  QUANT WIDTH LENGTH  QUANT WIDTH LE
   2NGTH TTLENG TRIM//(3(F7.0,F6.2,F7.2),F8.0,F4.0))
    WRITE(6,435)TOTLEN,AREA,SCRAP,AVTRIM,PCSCRP
435 FORMAT(71H LENGTH IN FT  AREA IN SQFT  SCRAP IN SQFT  AVER TRIM WI
   2DTH  PERC SCRAP//(3F14.0,2F14.2))
    WRITE(6,436) TOTSUT,TOTTIM,LISTB,LISTA
436 FORMAT(56H SET UP MINS  ELAPSED MINS  LIST B ORDERS  LIST A ORDERS
   2//(2F15.1,2I10))
```

```
      GTSUP=GTSUP+TOTSUT/60.
      GTTIM=GTTIM+TOTTIM/60.
      LATOT=LATOT+LISTA
      LBTOT=LBTOT+LISTB
      GTPWCH=GTPWCH+PWIDCH
      GTLENG=GTLENG+TOTLEN
      GTAREA=GTAREA+AREA
      GTSCRP=GTSCRP+SCRAP
      GTPCSP=GTSCRP*100./GTAREA
      GTAVTM=GTAREA*12./GTLENG
      KSTATM=KSTATM+1
  891 IF(KSTATM-40)440,440,450
  440 N=0
      GO TO (130,189,180,255,208,280,305,350,130,189,180,255,208,280,
     2305,350,130,189,180,255,208,280,305,350,55,105,120,155,30,199,
     3204,330,405,194,230,412,612,512,355,380),KSTATM
  450 WRITE(6,455)
  455 FORMAT(20X,33H SUMMARY DATA FOR PRECEDING CYCLE)
      WRITE(6,465)GTSUP,GTTIM
  465 FORMAT(42H TOTAL SET UP HOURS       TOTAL ELAPSED HOURS//(2F16.2))
      WRITE(6,470)GTLENG,GTAREA,GTAVTM,GTPCSP,GTPWCH
  470 FORMAT(70H TOT LENGTH   TOT AREA   AVER TRIM WIDTH   PERC SCRAP   NO O
     2F ROLL CHANGES////(2E14.6,3F12.2))
      WRITE(6,990) LATOT,LBTOT
  990 FORMAT(41H LIST A ORDER TOTAL      LIST B ORDER TOTAL//(2I18))
      JALLAG=JALLAG+1
      IF(JALLAG-JDOALL)5,5,460
  460 STOP
      END
$IBFTC GENR
      SUBROUTINE GEN(SIGQ,SIGW,SIGL,AVQ,AVW,AVL,RHO,Q,W,AL,WW,ALL,N,M,
     2WWYD)
      DIMENSION Q(200),W(200),AL(200),WW(200),ALL(200)
      M=N*9/10
      DO 30 I=1,N
      Q(I)=EXP(AVQ+SIGQ*GAURN(X))
      W(I)=EXP(AVW+SIGW*GAURN(X))
      IF(W(I)-77.25)15,10,10
   10 W(I)=77.25
   15 WIDTH=W(I)
   20 AL(I)=EXP(AVL+RHO*SIGL/SIGW*(ALOG(WIDTH)-AVW)+SIGL*SQRT(1.-RHO**2)
     2*GAURN(X))
      IF(AL(I)-150.)30,30,25
   25 AL(I)=150.
   30 CONTINUE
      IF(N-1)65,65,35
   35 DO 60 I=1,M
      WW(I)=EXP(AVW+SIGW*GAURN(X))
      IF(WW(I)-77.25)45,40,40
   40 WW(I)=77.25
   45 WIDTH=WW(I)
      ALL(I)=EXP(AVL+RHO*SIGL/SIGW*(ALOG(WIDTH)-AVW)+SIGL*SQRT(1.-RHO**2
     2)*GAURN(X))
      IF(ALL(I)-150.)60,60,55
   55 ALL(I)=150.
   60 CONTINUE
   65 RETURN
      END
$IBFTC SCHD
      SUBROUTINE SCHED(A,Q,QA,QB,QC,W,WA,WB,WC,AL,ALA,ALB,ALC,TWID,N,K,
     2TLENG,TOTSUT,WW,ALL,M,LISTB,WYD,WWYD,FLUT)
      DIMENSION Q(200),W(200),AL(200),TLEN(200),TWID(200),QA(200),
     2QB(200),QC(200),WA(200),WB(200),WC(200),ALA(200),ALB(200),
     3ALC(200),ANOUTW(200),QQ(200),WW(200),ALL(200),TLENG(200),
     4TWIT(200),KTWID(200),ALENG(200),BLENG(200),SUTIME(200),SUTIMA(200)
     5,SUTIMB(200),WAST(200)
      LISTB=0
      K=1
      A=WYD
      KSTAT=1
      BLONG=3000.
    3 DO 201 I=1,N
      IF(Q(I))201,201,4
    4 TLEN(I)=Q(I)*AL(I)/12.
      NOUT=WWYD/W(I)
      ANOUT=NOUT
      IF(N-1)55,55,6
    6 IF(ANOUT-5.)30,25,25
   25 ANOUT=4.
```

```
          GO TO 60
   30  IF(TLEN(I)/ANOUT-BLONG)35,40,40
   35  ANOUT=ANOUT-1.
       IF(ANOUT)38,38,30
   38  ANOUT=1.
       GO TO (201,40),KSTAT
   40  TWIT(I)=ANOUT*W(I)
       IF(TLEN(I)/ANOUT-BLONG)60,45,45
   45  IF(TWIT(I)-A)60,55,55
   60  DO 110 J=1,N
       IF(W(J))110,110,61
   61  IF(I-J)65,110,65
   65  TWIT(I)=ANOUT*W(I)+W(J)
       TLEN(I)=Q(I)*AL(I)/(12.*ANOUT)
       TLEN(J)=Q(J)*AL(J)/12.
       ANOTJ=1.
       IF(TWIT(I)-A)130,135,135
  100  TWIT(I)=W(I)+2.*W(J)
       TLEN(I)=Q(I)*AL(I)/12.
       TLEN(J)=Q(J)*AL(J)/24.
       ANOUT=1.
       ANOTJ=2.
       IF(TWIT(I)-A)105,135,135
  105  TWIT(I)=W(I)+3.*W(J)
       TLEN(J)=Q(J)*AL(J)/36.
       ANOUT=1.
       ANOTJ=3.
       IF(TWIT(I)-A)110,135,135
  130  TWIT(I)=W(I)+W(J)
       TLEN(I)=Q(I)*AL(I)/12.
       TLEN(J)=Q(J)*AL(J)/12.
       ANOUT=1.
       ANOTJ=1.
       IF(TWIT(I)-A)100,135,135
  135  IF(TWIT(I)-WWYD)140,140,110
  140  IF(TLEN(I)-TLEN(J))145,145,146
  146  GO TO (147,150),KSTAT
  147  IF(TLEN(J)-2000.)110,150,150
  145  IF(TLEN(I)-2000.)149,149,148
  148  PERC=.10
       GO TO 154
  149  PERC=.272/EXP(TLEN(I)/2000.)
  154  IF((1.+PERC/ANOUT)*TLEN(I)-(1.-PERC/ANOTJ)*TLEN(J))155,165,165
  150  IF(TLEN(J)-2000.)152,152,151
  151  PERC=.10
       GO TO 153
  152  PERC=.272/EXP(TLEN(J)/2000.)
  153  IF((1.-PERC/ANOUT)*TLEN(I)-(1.+PERC/ANOTJ)*TLEN(J))166,166,160
  155  W(I)=ANOUT*W(I)
       W(J)=ANOTJ*W(J)
       CALL CHECK(Q,W,AL,TWID,TLEN,TLENG,I,J,K,L,N,NO,ANOTL,ANOTJ,ANOUT,
      2WWYD)
       W(I)=W(I)/ANOUT
       W(J)=W(J)/ANOTJ
       IF(NO)110,110,115
  160  W(I)=ANOUT*W(I)
       W(J)=ANOTJ*W(J)
       CALL CHECK(Q,W,AL,TWID,TLEN,TLENG,J,I,K,L,N,NO,ANOTL,ANOUT,ANOTJ,
      2WWYD)
       W(I)=W(I)/ANOUT
       W(J)=W(J)/ANOTJ
       IF(NO)110,110,120
  110  CONTINUE
  201  CONTINUE
       GO TO 203
   55  TLENG(K)=TLEN(I)/ANOUT
       TWID(K)=ANOUT*W(I)
       QA(K)=Q(I)
       QB(K)=0.0
       QC(K)=0.0
       WA(K)=W(I)
       WB(K)=0.0
       WC(K)=0.0
       ALA(K)=AL(I)
       ALB(K)=0.0
       ALC(K)=0.0
       ALENG(K)=0.0
       BLENG(K)=0.0
       Q(I)=0.0
```

```
      W(I)=0.0
      AL(I)=0.0
      K=K+1
      GO TO 3
  165 TLENG(K)=TLEN(I)+ANOTJ/(ANOTJ+ANOUT)*(TLEN(J)-TLEN(I))
      GO TO 167
  166 TLENG(K)=TLEN(J)+ANOUT/(ANOTJ+ANOUT)*(TLEN(I)-TLEN(J))
  167 TWID(K)=TWIT(I)
      QA(K)=TLENG(K)*12.*ANOUT/AL(I)
      QB(K)=TLENG(K)*12.*ANOTJ/AL(J)
      QC(K)=0.0
      WA(K)=W(I)
      WB(K)=W(J)
      WC(K)=0.0
      ALA(K)=AL(I)
      ALB(K)=AL(J)
      ALC(K)=0.0
      ALENG(K)=0.0
      BLENG(K)=0.0
      Q(I)=0.C
      W(I)=0.0
      AL(I)=0.0
      Q(J)=0.0
      W(J)=0.0
      AL(J)=0.0
      K=K+1
      GO TO 3
  120 QA(K)=TLENG(K)*12.*ANOUT/AL(I)
      QB(K)=TLENG(K)*12.*ANOTJ/AL(J)*Q(J)/(Q(L)+Q(J))
      QC(K)=TLENG(K)*12.*ANOTL/AL(L)*Q(L)/(Q(L)+Q(J))
      ALENG(K)=QB(K)*AL(J)/(12.*ANOTJ)
      BLENG(K)=QC(K)*AL(L)/(12.*ANOTL)
      WA(K)=W(I)
      WB(K)=W(J)
      WC(K)=W(L)
      ALA(K)=AL(I)
      ALB(K)=AL(J)
      ALC(K)=AL(L)
      GO TO 125
  115 QA(K)=TLENG(K)*12.*ANOTJ/AL(J)
      QB(K)=TLENG(K)*12.*ANOUT/AL(I)*Q(I)/(Q(L)+Q(I))
      QC(K)=TLENG(K)*12.*ANOTL/AL(L)*Q(L)/(Q(L)+Q(I))
      ALENG(K)=QB(K)*AL(I)/(12.*ANOUT)
      BLENG(K)=QC(K)*AL(L)/(12.*ANOTL)
      WA(K)=W(J)
      WB(K)=W(I)
      WC(K)=W(L)
      ALA(K)=AL(J)
      ALB(K)=AL(I)
      ALC(K)=AL(L)
  125 Q(I)=0.0
      Q(J)=0.0
      Q(L)=0.0
      W(I)=0.0
      W(J)=0.0
      W(L)=0.0
      AL(I)=0.0
      AL(J)=0.0
      AL(L)=0.0
      K=K+1
      GO TO 3
  203 SUMQ=0.0
      DO 202 I=1,N
  202 SUMQ=SUMQ+Q(I)
      IF(SUMQ)247,205,205
  205 A=A-1.
      BLONG=BLONG-50.
      GO TO (206,208),KSTAT
  206 IF(A-WWYD+14.25)207,3,3
  207 KSTAT=KSTAT+1
      BLONG=2000.
      A=WYD
      GO TO 3        21.75
  208 IF(A-WWYD+22.00)209,3,3
  209 A=WYD
      BLONG=2000.
  213 DO 246 I=1,N
      IF(W(I))246,246,214
  214 TLEN(I)=Q(I)*AL(I)/12.
```

```
219 DO 245 J=1,M
    IF(W(I)+WW(J)-WWYD)220,245,245
220 IF(W(I)+WW(J)-A)226,225,225
225 IF(TLEN(I)-BLONG)245,228,228
228 QQ(J)=TLEN(I)*12./ALL(J)
    TWID(K)=W(I)+WW(J)
    TLENG(K)=TLEN(I)
    LISTB=LISTB+1
    GO TO 235
226 IF(2.*W(I)+WW(J)-WWYD)227,245,245
227 IF(2.*W(I)+WW(J)-A)245,230,230
230 IF(TLEN(I)/2.-BLONG)245,231,231
231 QQ(J)=TLEN(I)*12./(ALL(J)*2.)
    TLENG(K)=TLEN(I)/2.
    TWID(K)=2.*W(I)+WW(J)
    LISTB=LISTB+1
235 QA(K)=Q(I)
    QB(K)=QQ(J)
    QC(K)=0.0
    WA(K)=W(I)
    WB(K)=WW(J)
    WC(K)=0.0
    ALA(K)=AL(I)
    ALB(K)=ALL(J)
    ALC(K)=0.0
    ALENG(K)=0.0
    BLENG(K)=0.0
    Q(I)=0.0
    W(I)=0.0
    AL(I)=0.0
    QQ(J)=0.0
    WW(J)=0.0
    ALL(J)=0.0
    K=K+1
    GO TO 246
245 CONTINUE
246 CONTINUE
    SUMQ=0.0
    DO 402 I=1,N
402 SUMQ=SUMQ+Q(I)
    IF(SUMQ)247,247,347
347 A=A-1.
    BLONG=BLONG-50.
    IF(A-WWYD+17.25)350,213,213
350 A=WYD
319 DO 346 I=1,N
    IF(W(I))346,346,315
315 NOUT=WWYD/W(I)
    ANOUTW(I)=NOUT
    TLEN(I)=Q(I)*AL(I)/12.
    IF(ANOUTW(I)*W(I)-A)346,240,240
346 CONTINUE
    SUMQ=0.0
    DO 502 I=1,N
502 SUMQ=SUMQ+Q(I)
    IF(SUMQ)247,247,547
547 A=A-1.
    GO TO 319
240 TLENG(K)=TLEN(I)/ANOUTW(I)
    TWID(K)=ANOUTW(I)*W(I)
    QA(K)=Q(I)
    QB(K)=0.0
    QC(K)=0.0
    WA(K)=W(I)
    WB(K)=0.0
    WC(K)=0.0
    ALA(K)=AL(I)
    ALB(K)=0.0
    ALC(K)=0.0
    ALENG(K)=0.0
    BLENG(K)=0.0
    Q(I)=0.0
    W(I)=0.0
    AL(I)=0.0
    K=K+1
    GO TO 319
247 K=K-1
    TRIPTM=7.
    SWFPMK=315.
```

```
      SWFPMJ=185.
      DO 649 I=1,K
      SUTIME(I)=0.0
      SUTIMA(I)=0.0
  649 SUTIMB(I)=0.0
      IF(FLUT-2.)824,827,827
  824 DO 259 I=1,K
      IF(ALENG(I))250,250,252
  250 IF(TLENG(I)-TRIPTM*SWFPMK)251,256,256
  251 SUTIME(I)=(TRIPTM*SWFPMK-TLENG(I))/SWFPMK
      GO TO 259
  252 IF(ALENG(I)-TRIPTM*SWFPMK)253,257,257
  253 SUTIMA(I)=(TRIPTM*SWFPMK-ALENG(I))/SWFPMK
  359 IF(BLENG(I)-TRIPTM*SWFPMK)254,258,258
  254 SUTIMB(I)=(TRIPTM*SWFPMK-BLENG(I))/SWFPMK
  256 SUTIME(I)=0.0
      GO TO 259
  257 SUTIMA(I)=0.0
      GO TO 359
  258 SUTIMB(I)=0.0
  259 CONTINUE
      GO TO 469
  827 DO 459 I=1,K
      IF(ALENG(I))450,450,452
  450 IF(TLENG(I)-TRIPTM*SWFPMJ)451,456,456
  451 SUTIME(I)=(TRIPTM*SWFPMJ-TLENG(I))/SWFPMJ
      GO TO 459
  452 IF(ALENG(I)-TRIPTM*SWFPMJ)453,457,457
  453 SUTIMA(I)=(TRIPTM*SWFPMJ-ALENG(I))/SWFPMJ
  559 IF(BLENG(I)-TRIPTM*SWFPMJ)454,458,458
  454 SUTIMB(I)=(TRIPTM*SWFPMJ-BLENG(I))/SWFPMJ
  456 SUTIME(I)=0.0
      GO TO 459
  457 SUTIMA(I)=0.0
      GO TO 559
  458 SUTIMB(I)=0.0
  459 CONTINUE
  469 TOTSUT=0.0
      DO 261 I=1,K
  261 TOTSUT=TOTSUT+SUTIME(I)+SUTIMA(I) SUTIMB(I)
      DO 260 I=1,K
  275 KTWID(I)=TWID(I)+.73
      TWID(I)=KTWID(I)+1
  263 KX=TWID(I)/3.
      XX=KX
      X=TWID(I)/XX
      IF(X-3.)260,260,262
  262 TWID(I)=TWID(I)+1.
      GO TO 263
  260 CONTINUE
      IF(K-1)972,972,361
  361 MA=K-1
      DO 270 I=1,MA
      MB=K-1
      DO 270 J=1,MB
      IF(TWID(J)-TWID(J+1))265,270,270
  265 TEMPTW=TWID(J)
      TEMPTL=TLENG(J)
      TEMPQA=QA(J)
      TEMPQB=QB(J)
      TEMPQC=QC(J)
      TEMPWA=WA(J)
      TEMPWB=WB(J)
      TEMPWC=WC(J)
      TEMPLA=ALA(J)
      TEMPLB=ALB(J)
      TEMPLC=ALC(J)
      TWID(J)=TWID(J+1)
      TLENG(J)=TLENG(J+1)
      QA(J)=QA(J+1)
      QB(J)=QB(J+1)
      QC(J)=QC(J+1)
      WA(J)=WA(J+1)
      WB(J)=WB(J+1)
      WC(J)=WC(J+1)
      ALA(J)=ALA(J+1)
      ALB(J)=ALB(J+1)
      ALC(J)=ALC(J+1)
      TWID(J+1)=TEMPTW
```

```
      TLENG(J+1)=TEMPTL
      QA(J+1)=TEMPQA
      QB(J+1)=TEMPQB
      QC(J+1)=TEMPQC
      WA(J+1)=TEMPWA
      WB(J+1)=TEMPWB
      WC(J+1)=TEMPWC
      ALA(J+1)=TEMPLA
      ALB(J+1)=TEMPLB
      ALC(J+1)=TEMPLC
  270 CONTINUE
      KK=K-1
      DO 960 I=1,KK
      L=K-I+1
      IF(TWID(L-1)-TWID(L))956,956,955
  955 WAST(L)=(TWID(L-1)-TWID(L))*TLENG(L)/12.
      GO TO 960
  956 WAST(L)=0.0
  960 CONTINUE
      L=K
  961 IF(WAST(L))964,964,965
  965 IF(WAST(L)-2500.)966,966,964
  966 TWID(L)=TWID(L-1)
      IF(L-2)972,972,967
  964 L=L-1
  970 IF(L-1)972,972,961
  967 IF(TWID(L-2)-TWID(L))968,968,969
  969 L=L-2
      GO TO 970
  968 IF(L-3)972,972,973
  973 IF(TWID(L-3)-TWID(L))974,974,975
  975 L=L-3
      GO TO 970
  974 IF(L-4)972,972,976
  976 IF(TWID(L-4)-TWID(L))977,977,978
  978 L=L-4
      GO TO 970
  977 IF(L-5)972,972,979
  979 IF(TWID(L-5)-TWID(L))980,980,981
  981 L=L-5
      GO TO 970
  980 IF(L-6)972,972,981
  972 RETURN
      END
$IBFTC CHCK
      SUBROUTINE CHECK(Q,W,AL,TWID,TLEN,TLENG,IA,JA,K,L,N,NO,ANOTL,
     2ANOTJ,ANOUT,WWYD)
      DIMENSION Q(200),W(200),AL(200),TWID(200),TLEN(200),TLENG(200)
      NO=0
      DO 80 L=1,N
      IF(IA-L)5,80,5
    5 IF(JA-L)10,80,10
   10 DAW=W(L)-W(IA)
      WAA=W(L)+W(JA)
      WAB=W(IA)+W(JA)
      ANOTL=1.
      TLEN(L)=Q(L)*AL(L)/12.
      TLEN(IA)=Q(IA)*AL(IA)/(12.*ANOUT)
      TLEN(JA)=Q(JA)*AL(JA)/(12.*ANOTJ)
      IF(TLEN(JA)-2000.)11,12,12
   11 AAW=4.-(2000.-TLEN(JA))/1000.
      GO TO 14
   12 AAW=2.
   14 IF(DAW)20,20,15
   15 IF(DAW-AAW)60,60,80
   20 IF(DAW+AAW)25,60,60
   25 DBW=2.*W(L)-W(IA)
      WAA=2.*W(L)+W(JA)
      WAB=W(IA)+W(JA)
      ANOTL=2.
      TLEN(L)=Q(L)*AL(L)/24.
      IF(DBW)35,35,30
   30 IF(DBW-AAW)60,60,80
   35 IF(DBW+AAW)40,60,60
   40 DCW=3.*W(L)-W(IA)
      WAA=3.*W(L)+W(JA)
      WAB=W(IA)+W(JA)
      ANOTL=3.
      TLEN(L)=Q(L)*AL(L)/36.
```

```
      IF(DCW)50,50,45
   45 IF(DCW-AAW)60,60,80
   50 IF(DCW+AAW)80,60,60
   60 IF(TLEN(L)+TLEN(IA)-TLEN(JA))65,65,70
   65 IF(TLEN(L)+TLEN(IA)-2000.)67,67,66
   66 PER=.10
      GO TO 68
   67 PER=.272/EXP((TLEN(L)+TLEN(IA))/20  .)
   68 IF((1.+PER/ANOTL)*TLEN(L)+(1.+PER/ANOUT)*TLEN(IA)-(1.-PER/ANOTJ)
     2*TLEN(JA))80,73,73
   70 IF(TLEN(JA)-2000.)172,172,171
  171 PER=.10
      GO TO 173
  172 PER=.272/EXP(TLEN(IA)/2000.)
  173 IF((1.-PER/ANOTL)*TLEN(L)+(1.-PER/ANOUT)*TLEN(IA)-(1.+PER/ANOTJ)
     2*TLEN(JA))74,74,80
   73 X=(TLEN(JA)/ANOTJ-TLEN(L)/ANOTL-TLEN(IA)/ANOUT)/(TLEN(JA)/ANOTJ**
     2+TLEN(L)/ANOTL**2+TLEN(IA)/ANOUT**2)
      TLENG(K)=(1.-X/ANOTJ)*TLEN(JA)
      GO TO 72
   74 X=(TLEN(L)/ANOTL+TLEN(IA)/ANOUT-TLEN(JA)/ANOTJ)/(TLEN(L)/ANOTL**2
     2+TLEN(IA)/ANOUT**2+TLEN(JA)/ANOTJ**2)
      TLENG(K)=(1.+X/ANOTJ)*TLEN(JA)
   72 IF(WAA-WAB)76,76,77
   76 TWID(K)=WAB
      GO TO 78
   77 TWID(K)=WAA
   78 IF(TWID(K)-WWYD)79,79,80
   79 NO=1
      GO TO 85
   80 CONTINUE
   85 RETURN
      END
$IBMAP RANDPK
       ENTRY    EXPRN
       ENTRY    GAURN
       ENTRY    FLRAN
       ENTRY    GETNM
       ENTRY    STORNM
 EXPRN LDQ      RANDOM
 C     PXD      952,0
 H     STA      A
       MPY      GENERA
       STQ      COMMON+1
       STQ      COMMON
 F     MPY      GENERA
       STQ      RANDOM
       CLA      COMMON
       TLQ      B
       LDQ      COMMON+1
       RQL      12
       CAL      C
       LGL      24
       STO      COMMON
       CLA      A
       LLS      12
 E     FAD      COMMON
 G     TNZ      1,4
       TRA      E
 B     MPY      GENERA
       STQ      COMMON
       CLA      RANDOM
       TLQ      F
       CLA      A
       ADM      G
       TRA      H
 GAURN SXD      COMMON+3,4
 CC    TSX      EXPRN,4
       ADD      AA
       STO      COMMON+4
       TSX      EXPRN,4
       STO      COMMON
       FSB      BB
       STO      COMMON+1
       LDQ      COMMON+1
       FMP      COMMON+1
       SUB      COMMON+4
       TPL      CC
       LXD      COMMON+3,4
```

```
         CLA    COMMON
S        LDQ    RANDOM
         RQL    20
         LLS    0
         TRA    1,4
FLRAN    LDQ    RANDOM
         MPY    GENERA
         STQ    RANDOM
         CLA    AAA
         LGL    28
         FAD    AAA
         TRA    S
GETNM    CLA    RANDOM
         STO*   3,4
         TRA    1,4
STORNM   CLA*   3,4
         STO    RANDOM
         TRA    1,4
GENERA   OCT    343277244615
RANDOM   DEC    30517578125
AA       OCT    001000000000
BB       DEC    1.
AAA      OCT    172000000100
A        OCT    00021700000
COMMON   BSS    5
         END
$DATA
```

67	64	78	16	6	7625	7725
434	387	433	6456	3478	4129	3
1141	436	244	8047	3482	3911	365
1336	448	266	7435	3127	4002	70
1385	406	182	7137	3489	3725	-236
947	356	354	7819	3150	4060	693
1097	412	260	7853	3379	4047	521
999	251	126	7110	3106	3889	- 389
606	361	451	6707	3810	4434	631
1017	348	130	7638	3269	4143	- 318
808	626	285	6928	3732	4565	905
821	233	125	8289	4158	4662	1000
328	167	157	7745	3365	3985	- 378
472	162	111	7630	3225	4132	548
1083	571	164	6765	3229	3941	194
676	432	249	6951	3570	3687	4
785	266	380	8037	3378	3895	- 630
604	130	169	7722	3105	3774	- 889
272	000	0	8762	3829	3761	1000
1117	536	183	7295	3541	3850	- 184
745	392	603	6552	3410	4195	185
469	65	49	7001	4020	4504	- 30
450	11	000	7794	3792	4727	000
832	213	440	6555	3866	4550	966
1033	455	187	6637	2734	3918	- 747
1313	528	172	6760	3468	3802	- 267

BIBLIOGRAPHY

Books

Aitcheson, J. and Brown, J.A.C. The Lognormal Distribution. Cambridge:
University Press, 1957.

Alchian, Armen. "Costs and Outputs," The Allocation of Economic
Resources. Stanford, California: Stanford University Press, 1959.

Anthony, R. N. Management Accounting, Homewood, Illinois: Richard
D. Irwin, Inc., Fourth Edition, 1970.

Aris, R. The Optimal Design of Chemical Reactors. New York: Academic
Press, 1961.

Asher, Harold. Cost Quantity Relationship in the Airframe Industry.
Santa Monica, California: The Rand Corporation, 1956.

Bierman, Harold Jr., L. E. Fouraker, and R. K. Jaedicke. Quantitative
Analysis for Business Decisions. Homewood, Illinois: Richard D. Irwin,
Inc., 1967.

Bowman, E. H., and R. B. Fetter. Analysis for Production and Operations
Management. Third Edition, Homewood, Illinois: Richard D. Irwin, Inc.,
1967.

Brown, R. G. Smoothing, Forecasting and Prediction of Discrete Time
Series. Englewood Cliffs, N. J.: Prentice-Hall, Inc., 1963.

_____. Statistical Forecasting for Inventory Control. New York:
McGraw-Hill, 1959.

Carr, C. R. and C. W. Howe. Quantitative Decision Procedures in Management and Economics. New York: McGraw-Hill, 1964.

Chilton, C. H., (ed.). Cost Engineering in the Process Industries. New York:
McGraw-Hill, 1960.

Davidson, R. K., V. L. Smith, and J. W. Wiley. Economics: An Analytical
Approach. Revised Edition. Homewood, Illinois: Richard D. Irwin, Inc.,
1962.

Eshbach, O. W., (ed.). Handbook of Engineering Fundamentals. Second
Edition. New York: John Wiley and Sons, 1958, pp. 8.15-8.19.

467

Freund, J.E. Mathematical Statistics. Englewood Cliffs, N.J.: Prentice Hall, Inc., 1962.

Goldberg, Samuel. Introduction to Difference Equations. New York: John Wiley and Sons, Inc., Science Editions, 1961.

Grayson, C.J. Jr., Decisions Under Uncertainty. Boston: Harvard University, Graduate School of Business Administration, 1960.

Guibert, Paul. Le Plan de Fabrication Aeronautique. Paris: Dunod, 1945. English translation under the title, Mathematical Studies of Aircraft Construction, is available from Central Air Documents Office, Wright-Patterson Air Force Base, Dayton, Ohio.

Holt, C.C., Franco Modigliani, J. F. Muth and H. A. Simon. Planning Production, Inventories, and Work Force. Englewood Cliffs, N.J.: Prentice-Hall, Inc., 1960.

Leontief, Wassily. The Structure of the American Economy, 1919-1939. Second Edition. New York: Oxford University Press, 1951.

Lutz, Friedrich and Vera. The Theory of Investment of the Firm. Princeton: Princeton University Press, 1951.

Magee, J.F. Production Planning and Inventory Control. New York: McGraw-Hill, 1958.

Manne, A.S. Investments for Capacity Expansion. Cambridge, Mass.: The M.I.T. Press, 1967.

McMillan, Claude and R. F. Gonzalez. Systems Analysis. Homewood, Illinois: Richard D. Irwin, Inc., 1968.

Meier, R.C., W. T. Newell and H. L. Pazer. Simulation in Business and Economics. Englewood Cliffs, N.J.: Prentice-Hall, Inc., 1969.

Mize, J.H. and J. G. Cox. Essentials of Simulation. Englewood Cliffs, N.J.: Prentice-Hall, Inc., 1968.

Mood, A.M. and F. A. Graybill. Introduction to the Theory of Statistics. New York: McGraw-Hill, 1963.

Nicholls, W.H. Labor Productivity Functions in Meat Packing. Chicago: Chicago University Press, 1948.

Pearson, E.S. and H. O. Hartley. Biometrika Tables for Statisticians. Vol. I, Cambridge: University Press, 1958.

Raiffa, Howard. Decision Analysis. Reading, Massachusetts: Addison-Wesley, 1968.

468

Raiffa, Howard and Robert Schlaifer. *Applied Statistical Decision Theory*. Boston, Massachusetts: Harvard University, Division of Research, Graduate School of Business Administration, 1961.

Roberts, S.M. *Dynamic Programming in Chemical Engineering and Process Control*. New York: Academic Press, 1964.

Schlaifer, Robert. *Probability and Statistics for Business Decisions*. New York: McGraw-Hill Book Company, 1959.

Schrieber, A. N. *Corporate Simulation Models*, Graduate School of Business Administration, University of Washington, 1970.

Schweyer, H.E. *Process Engineering Economics*. New York: McGraw-Hill, 1955.

Smith, V.L. *Investment and Production*. Cambridge, Massachusetts: Harvard University Press, 1961.

Spencer, M.H. *Managerial Economics*. Third Edition. Homewood, Illinois: Richard D. Irwin, Inc., 1968.

Terborgh, George. *Dynamic Equipment Policy*. New York: McGraw-Hill, 1949.

Tocher, K.D. *The Art of Simulation*. Princeton, New Jersey: D. Van Nostrand Co., Inc., 1963.

Vilbrandt, F.C. and C. E. Dryden. *Chemical Engineering Plant Design*. Fourth Edition. New York: McGraw-Hill, 1959.

Articles

Armour, G. C., and E. S. Buffa. "A Heuristic Algorithm and Computer Simulation Approach to the Relative Location of Facilities," Management Science, Vol. 9, No. 2 (January, 1963), pp. 294-309.

Baldwin, R. H. "How to Assess Investment Proposals," Harvard Business Review, May-June 1959, pp. 98-99.

Baloff, Nicholas. "Estimating the Parameters of the Startup Model - An Empirical Approach," The Journal of Industrial Engineering, Vol. 18, No. 4 (April, 1967), pp. 248-253.

Bass, F. M. "Marketing Research Expenditures - A Decision Model," The Journal of Business, Vol. 26 (January, 1963), pp. 77-90.

Bellman, Richard. "Top Management Decision and Simulation Processes," Journal of Industrial Engineering, Vol. 9 (September-October, 1958), p. 461.

Brown, R. V. "Do Managers Find Decision Theory Useful?" Harvard Business Review, May-June, 1970, p. 78.

Buffa, E.S., G.C. Armour, and T. E. Vollman. "Allocating Facilities with CRAFT," Harvard Business Review, (March-April, 1964), pp. 136-58.

Cameron, D. C. "Travel Charts - A Tool for Analyzing Material Movement Problems," Modern Materials Handling, Volume 8, No. 1.

Carr, G. W. "Peacetime Cost Estimating Requires New Learning Curves," Aviation, Vol. 45 (April, 1946), pp. 76-77.

Carter, E. C. "What are the Risks in Risk Analysis," Harvard Business Review, July-August 1972, p. 72.

Chambers, J.C., S. K. Mullick, and D. D. Smith. "How to Choose the Right Forecasting Technique," Harvard Business Review, July-August, 1971, p. 45.

Chenery, B. "Engineering Production Functions," Quarterly Journal of Economics, Vol. 63 (November, 1949), pp. 507-31.

Cobb, C. W. and P.H. Douglas. "A Theory of Production," American Economic Review, Supplement, (March, 1928), pp. 139-65.

DeJong, J.R. "The Effects of Increasing Skills on Cycle Time and Its Consequences for Time Standards," Ergonomics, Vol. 1, No. 1 (1957), pp. 51-59.

Frankfurter, G.M., K. E. Kendall, C. C. Pegels and P.D. Wharton. "Management Control of Blood through a Short Term Supply-Demand Forecast System," School of Management, State University of New York at Buffalo, 1971.

Hanssmann, Fred and S. W. Hess. "A Linear Programming Approach to Production and Employment Scheduling," Management Technology, (January, 1960), pp. 46-51.

Heady, O. "An Econometric Investigation of the Technology of Agricultural Production Functions," Econometrica, Vol. 25 (April, 1957), pp. 249-68.

Hertz, D. B. "Risk Analysis in Capital Investment," Harvard Business Review, (January-February, 1964), pp. 95-106.

Hess, S. W. and H. A. Quigley. "Analysis of Risk in Investments Using Monte Carlo Techniques," Chemical Engineering Progress Symposium Series 42: Statistics and Numerical Methods in Chemical Engineering, New York: American Institute of Chemical Engineering, (1963), p. 55.

Hirschleifer, Jack. "The Bayesian Approach to Statistical Decision – An Exposition," The Journal of Business, Vol. 24 (October, 1961), pp. 471-89.

_____. "The Firm's Cost Function: A Successful Reconstruction," The Journal of Business, Vol. 35 (July, 1962), pp. 235-55.

Jen, F.C., C.C. Pegels, and J. M. Dupuis. "Optimal Capacities of Production Facilities," Management Science, Vol. 14, No. 10 (June, 1968), pp. B 573-80.

Katell, S. and J. H. Faber. "What Does Tonnage Oxygen Cost?" Chemical Engineering, (June 29, 1959), pp. 107-10).

Levy, F. K. "Adaption in the Production Process," Management Science, Vol. 11, No. 6 (April, 1965), pp. 136-54.

Magee, J.F. "How to Use Decision Trees in Capital Investment," Harvard Business Review, (September-October, 1964), pp. 79-96.

Mitten, L. G. and G. L. Nemhauser. "Multistage Optimization," Chemical Engineering Progress, Vol. 59, No. 1 (January, 1963), pp. 52-60.

_____. "Optimization of Multistage Separation Processes by Dynamic Programming," The Canadian Journal of Chemical Engineering, (October, 1963), pp. 187-94.

Orr, Daniel. "Costs and Outputs: An Appraisal of Dynamic Aspects," The Journal of Business, Vo. 37 (January, 1964), pp. 51-60.

471

Parker, G. C. and E. L. Segura, "How to Get a Better Forecast," *Harvard Business Review*, March-April 1971, Vol. 49-2, p. 99.

Pegels, C. C. "Plant Layout and Discrete Optimizing," *The International Journal of Production Research*, Vol. 5, No. 1 (March, 1966), pp. 81-92.

_____. "A Technique for Determining Optimal Machine Characteristics," *The International Journal of Production Research*, Vol. 6, No. 1 (1967), pp. 47-56.

_____. "A Management Science Approach to Power Factor Correction," *The International Journal of Electrical Engineering Education*, Vol. 5 (1967), pp. 691-97.

_____. "A Comparison of Decision Criteria for Capital Investment Decisions," *The Engineering Economist*, Vol. 13 (September, 1968), pp. 211-20.

_____. "Simulation and the Optimal Design of a Production Process," *The International Journal of Production Research*, Vol. 7, No. 3 (1969), pp. 219-31.

_____. "Work Force Planning for the Job Shop," *Logistics Review*, Vol. 5, No. 21 (1969), pp. 41-48.

_____. "Exponential Forecasting - Some New Variations," *Management Science*, Vol. 15, No. 5 (January, 1969), pp. 311-15.

_____. "On Startup or Learning Curves: An Expanded View," *AIIE Transactions*, Vol. 1, No. 2 (September, 1969), pp. 216-22.

Reiter, Stanley, and G. R. Sherman. "Discrete Optimizing," *SIAM Journal*, Vol. 13, No. 3 (1965), pp. 864-89.

Shubik, Martin. "Simulation of the Industry and the Firm," *American Economic Review*, Vol. 50 (December, 1960), p. 909.

Smith, S.B. "An Input-Output Model for Production and Inventory Planning," *Journal of Industrial Engineering*, (January-February, 1965), pp. 64-69.

Tintner, George. "A Note on the Derivation of Production Functions from Farm Records Data," *Econometrica*, Vol. 12, No. 1 (January, 1944), pp. 26-34.

Vandell, R. F., "Management Evolution in the Quantitative World," *Harvard Business Review*, January-February, 1970.

Wright, T.P. "Factors Affecting the Cost of Airplanes," *Journal of the Aeronautical Sciences*, Vol. 3 (February, 1936), pp. 122-28.